HARCOURT
Science

Harcourt School Publishers

Orlando • Boston • Dallas • Chicago • San Diego

www.harcourtschool.com

Cover Image
This butterfly is a Red Cracker. It is almost completely red on its underside. It is called a cracker because the males make a crackling sound as they fly. The Red Cracker is found in Central and South America.

Copyright © 2000 by Harcourt, Inc.

All rights reserved. No part of this publication may be reproduced or transmitted in any form or by any means, electronic or mechanical, including photocopy, recording, or any information storage and retrieval system, without permission in writing from the publisher.

Requests for permission to make copies of any part of the work should be mailed to the following address:

School Permissions, Harcourt, Inc.
6277 Sea Harbor Drive
Orlando, FL 32887-6777

HARCOURT and the Harcourt Logo are trademarks of Harcourt, Inc.

sciLINKS is owned and provided by the National Science Teachers Association. All rights reserved.

Smithsonian Institution Internet Connections owned and provided by the Smithsonian Institution. All other material owned and provided by Harcourt School Publishers under copyright appearing above.

The name of the Smithsonian Institution and the Sunburst logo are registered trademarks of the Smithsonian Institution. The copyright in the Smithsonian website and Smithsonian website pages are owned by the Smithsonian Institution.

Printed in the United States of America

ISBN 0-15-315686-4	UNIT A
ISBN 0-15-315687-2	UNIT B
ISBN 0-15-315688-0	UNIT C
ISBN 0-15-315689-9	UNIT D
ISBN 0-15-315690-2	UNIT E
ISBN 0-15-315691-0	UNIT F

2 3 4 5 6 7 8 9 10   032   2000

Authors

Marjorie Slavick Frank
Former Adjunct Faculty Member at Hunter, Brooklyn, and Manhattan Colleges
New York, New York

Robert M. Jones
Professor of Education
University of Houston-Clear Lake
Houston, Texas

Gerald H. Krockover
Professor of Earth and Atmospheric Science Education
School Mathematics and Science Center
Purdue University
West Lafayette, Indiana

Mozell P. Lang
Science Education Consultant
Michigan Department of Education
Lansing, Michigan

Joyce C. McLeod
Visiting Professor
Rollins College
Winter Park, Florida

Carol J. Valenta
Vice President—Education, Exhibits, and Programs
St. Louis Science Center
St. Louis, Missouri

Barry A. Van Deman
Science Program Director
Arlington, Virginia

Senior Editorial Advisor

Napoleon Adebola Bryant, Jr.
Professor Emeritus of Education
Xavier University
Cincinnati, Ohio

Program Advisors

Michael J. Bell
Assistant Professor of Early
 Childhood Education
School of Education
University of Houston-Clear Lake
Houston, Texas

George W. Bright
Professor of Mathematics Education
The University of North Carolina at
 Greensboro
Greensboro, North Carolina

Pansy Cowder
Science Specialist
Tampa, Florida

Nancy Dobbs
Science Specialist, Heflin Elementary
Alief ISD
Houston, Texas

Robert H. Fronk
Head, Science/Mathematics
 Education Department
Florida Institute of Technology
Melbourne, Florida

Gloria R. Guerrero
Education Consultant
Specialist in English as a Second
 Language
San Antonio, Texas

Bernard A. Harris, Jr.
Physician and Former Astronaut
(STS 55—Space Shuttle Columbia,
STS 63—Space Shuttle Discovery)
Vice President, SPACEHAB Inc.
Houston, Texas

Lois Harrison-Jones
Education and Management
 Consultant
Dallas, Texas

Linda Levine
Educational Consultant
Orlando, Florida

Bertie Lopez
Curriculum and Support Specialist
Ysleta ISD
El Paso, Texas

Kenneth R. Mechling
Professor of Biology and Science
 Education
Clarion University of Pennsylvania
Clarion, Pennsylvania

Nancy Roser
Professor of Language and Literacy
 Studies
University of Texas, Austin
Austin, Texas

Program Advisor and Activities Writer

Barbara ten Brink
Science Director
Round Rock Independent School
 District
Round Rock, Texas

Reviewers and Contributors

Dorothy J. Finnell
Curriculum Consultant
Houston, Texas

Kathy Harkness
Retired Teacher
Brea, California

Roberta W. Hudgins
Teacher, W. T. Moore Elementary
Tallahassee, Florida

Libby Laughlin
Teacher, North Hill Elementary
Burlington, Iowa

Teresa McMillan
Teacher-in-Residence
University of Houston-Clear Lake
Houston, Texas

Kari A. Miller
Teacher, Dover Elementary
Dover, Pennsylvania

Julie Robinson
Science Specialist, K-5
Ben Franklin Science Academy
Muskogee, Oklahoma

Michael F. Ryan
Educational Technology Specialist
Lake County Schools
Tavares, Florida

Judy Taylor
Teacher, Silvestri Junior High
 School
Las Vegas, Nevada

UNIT A

LIFE SCIENCE
Plants and Animals

Chapter 1	**How Plants Grow**	**A2**
	Lesson 1—What Do Plants Need?	A4
	Lesson 2—What Do Seeds Do?	A10
	Lesson 3—How Do Plants Make Food?	A18
	Science and Technology • Drought-Resistant Plants	A24
	People in Science • George Washington Carver	A26
	Activities for Home or School	A27
	Chapter Review and Test Preparation	A28

Chapter 2	**Types of Animals**	**A30**
	Lesson 1—What Is an Animal?	A32
	Lesson 2—What Are Mammals and Birds?	A40
	Lesson 3—What Are Amphibians, Fish, and Reptiles?	A48
	Science Through Time • Discovering Animals	A58
	People in Science • Rodolfo Dirzo	A60
	Activities for Home or School	A61
	Chapter Review and Test Preparation	A62

Unit Project Wrap Up   A64

UNIT B

LIFE SCIENCE
Plants and Animals Interact

Chapter 1

Where Living Things are Found — **B2**

Lesson 1—What Are Ecosystems? — B4
Lesson 2—What Are Forest Ecosystems? — B10
Lesson 3—What Is a Desert Ecosystem? — B18
Lesson 4—What Are Water Ecosystems? — B24
   Science and Technology • Using Computers to
                         Describe the Environment — B32
   People in Science • Margaret Morse Nice — B34
   Activities for Home or School — B35
Chapter Review and Test Preparation — B36

Chapter 2

Living Things Depend on One Another — **B38**

Lesson 1—How Do Animals Get Food? — B40
Lesson 2—What Are Food Chains? — B46
Lesson 3—What Are Food Webs? — B52
   Science Through Time • People and Animals — B58
   People in Science • Akira Akubo — B60
   Activities for Home or School — B61
Chapter Review and Test Preparation — B62

Unit Project Wrap Up — B64

UNIT C

EARTH SCIENCE
Earth's Land

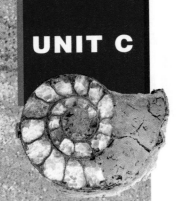

Chapter 1	**Rocks, Minerals, and Fossils**	**C2**
	Lesson 1—What Are Minerals and Rocks?	C4
	Lesson 2—How Do Rocks Form?	C10
	Lesson 3—What Are Fossils?	C18
	Science Through Time • Discovering Dinosaurs	C24
	People in Science • Charles Langmuir	C26
	Activities for Home or School	C27
	Chapter Review and Test Preparation	C28
Chapter 2	**Forces That Shape the Land**	**C30**
	Lesson 1—What Are Landforms?	C32
	Lesson 2—What Are Slow Landform Changes?	C38
	Lesson 3—What Are Rapid Landform Changes?	C46
	Science and Technology • Earthquake-Proof Buildings	C52
	People in Science • Scott Rowland	C54
	Activities for Home or School	C55
	Chapter Review and Test Preparation	C56
Chapter 3	**Soils**	**C58**
	Lesson 1—How Do Soils Form?	C60
	Lesson 2—How Do Soils Differ?	C66
	Lesson 3—How Can People Conserve Soil?	C72
	Science and Technology • Farming with GPS	C78
	People in Science • Diana Wall	C80
	Activities for Home or School	C81
	Chapter Review and Test Preparation	C82
Chapter 4	**Earth's Resources**	**C84**
	Lesson 1— What Are Resources?	C86
	Lesson 2— What Are Different Kinds of Resources?	C92
	Lesson 3— How Can We Conserve Earth's Resources?	C98
	Science and Technology • Recycling Plastic to Make Clothing	C106
	People in Science • Marisa Quinones	C108
	Activities for Home or School	C109
	Chapter Review and Test Preparation	C110
	Unit Project Wrap Up	C112

UNIT D

EARTH SCIENCE
Cycles on Earth and In Space

Chapter 1	**The Water Cycle**	**D2**
Lesson 1—Where Is Water Found on Earth?	D4
Lesson 2—What Is the Water Cycle?	D14
Science and Technology • A Filter for Clean Water	D20
People in Science • Lisa Rossbacher	D22
Activities for Home or School	D23
Chapter Review and Test Preparation	D24

Chapter 2	**Observing Weather**	**D26**
Lesson 1—What Is Weather?	D28
Lesson 2—How Are Weather Conditions Measured?	D34
Lesson 3—What Is a Weather Map?	D42
Science and Technology • Controlling Lightning Strikes	D48
People in Science • June Bacon-Bercey	D50
Activities for Home or School	D51
Chapter Review and Test Preparation	D52

Chapter 3	**Earth and Its Place in the Solar System**	**D54**
Lesson 1—What Is the Solar System?	D56
Lesson 2—What Causes Earth's Seasons?	D66
Lesson 3—How Do the Moon and Earth Interact?	D74
Lesson 4—What Is Beyond the Solar System?	D82
Science Through Time • Sky Watchers	D90
People in Science • Mae C. Jemison	D92
Activities for Home or School	D93
Chapter Review and Test Preparation	D94

Unit Project Wrap Up   D96

UNIT E

PHYSICAL SCIENCE
Investigating Matter

Chapter 1 — Properties of Matter — E2
Lesson 1—What Are Physical Properties of Matter? — E4
Lesson 2—What Are Solids, Liquids, and Gases? — E14
Lesson 3—How Can Matter Be Measured? — E20
   Science Through Time • Classifying Matter — E30
   People in Science • Dorothy Crowfoot Hodgkin — E32
   Activities for Home or School — E33
Chapter Review and Test Preparation — E34

Chapter 2 — Changes in Matter — E36
Lesson 1—What Are Physical Changes? — E38
Lesson 2—What Are Chemical Changes? — E44
   Science and Technology • Plastic Bridges — E50
   People in Science • Enrico Fermi — E52
   Activities for Home or School — E53
Chapter Review and Test Preparation — E54

Unit Project Wrap Up — E56

UNIT F

PHYSICAL SCIENCE
Exploring Energy and Forces

Chapter 1 — Heat — F2
Lesson 1—What Is Heat? — F4
Lesson 2—How Does Thermal Energy Move? — F12
Lesson 3—How Is Temperature Measured? — F18
   Science and Technology • Technology Delivers Hot Pizza — F24
   People in Science • Percy Spencer — F26
   Activities for Home or School — F27
Chapter Review and Test Preparation — F28

Chapter 2 — Light — F30
Lesson 1—How Does Light Behave? — F32
Lesson 2—How Are Light and Color Related? — F42
   Science Through Time • Discovering Light and Optics — F48
   People in Science • Lewis Howard Latimer — F50
   Activities for Home or School — F51
Chapter Review and Test Preparation — F52

Chapter 3 — Forces and Motion — F54
Lesson 1— How Do Forces Cause Motion? — F56
Lesson 2—What Is Work? — F64
Lesson 3—What Are Simple Machines? — F68
   Science and Technology • Programmable Toy Building Bricks — F74
   People in Science • Christine Darden — F76
   Activities for Home or School — F77
Chapter Review and Test Preparation — F78

Unit Project Wrap Up — F80

References — R1
   Science Handbook — R2
   Health Handbook — R11
   Glossary — R46
   Index — R54

Using Science Process Skills

When scientists try to find an answer to a question or do an experiment, they use thinking tools called process skills. You use many of the process skills whenever you think, listen, read, and write. Think about how these students used process skills to help them answer questions and do experiments.

Maria is interested in birds. She carefully observes the birds she finds. Then she uses her book to identify the birds and learn more about them.

Try This Find something outdoors that you want to learn more about. Use your senses to observe it carefully.

Talk About It What senses does Maria use to observe the birds?

Process Skills

Observe — use your senses to learn about objects and events

Charles finds rocks for a rock collection. He observes the rocks he finds. He compares their colors, shapes, sizes, and textures. He classifies them into groups according to their colors.

Try This Use the skills of comparing and classifying to organize a collection of objects.

Talk About It What other ways can Charles classify the rocks in his collection?

Process Skills

Compare — identify characteristics of things or events to find out how they are alike and different

Classify — group or organize objects or events in categories based on specific characteristics

Katie measures her plants to see how they grow from day to day. Each day after she **measures** she **records the data**. Recording the data will let her work with it later. She **displays the data** in a graph.

Try This Find a shadow in your room. Measure its length each hour. Record your data, and find a way to display it.

Talk About It How does displaying your data help you communicate with others?

Process Skills

Measure — compare mass, length, or capacity of an object to a unit, such as gram, centimeter, or liter

Record Data — write down observations

Display Data — make tables, charts, or graphs

An ad about low-fat potato chips claims that low-fat chips have half the fat of regular potato chips. Tani **plans and conducts an investigation** to test the claim.

Tani labels a paper bag Regular and Low-Fat. He finds two chips of each kind that are the same size, and places them above their labels. He crushes all the chips flat against the bag. He sets the stopwatch for one hour.

Tani **predicts** that regular chips will make larger grease spots on the bag than low-fat chips. When the stopwatch signals, he checks the spots. The spots above the Regular label are larger than the spots above the Low-Fat label. Tani **infers** that the claim is correct.

Try This Plan and conduct an investigation to test claims for a product. Make a prediction, and tell what you infer from the results.

Talk About It Why did Tani test potato chips of the same size?

Process Skills

Plan and conduct investigations—identify and perform the steps necessary to find the answer to a question

Predict—form an idea of an expected outcome based on observations or experience

Infer—use logical reasoning to explain events and make conclusions

You will have many opportunities to practice and apply these and other process skills in *Harcourt Science*. An exciting year of science discoveries lies ahead!

Safety in Science

Here are some safety rules to follow.

1. **Think ahead.** Study the steps and safety symbols of the investigation so you know what to expect. If you have any questions, ask your teacher.

2. **Be neat.** Keep your work area clean. If you have long hair, pull it back so it doesn't get in the way. Roll up long sleeves. If you should spill or break something, or get cut, tell your teacher right away.

3. **Watch your eyes.** Wear safety goggles when told to do so.

4. **Yuck!** Never eat or drink anything during a science activity unless you are told to do so by your teacher.

5. **Don't get shocked.** Be sure that electric cords are in a safe place where you can't trip over them. Don't ever pull a plug out of an outlet by pulling on the cord.

6. **Keep it clean.** Always clean up when you have finished. Put everything away and wash your hands.

In some activities you will see these symbols. They are signs for what you need to do to be safe.

Be especially careful.

Wear safety goggles.

Be careful with sharp objects.

Don't get burned.

Protect your clothes.

Protect your hands with mitts.

Be careful with electricity.

UNIT B

LIFE SCIENCE

Plants and Animals Interact

Chapter 1   **Where Living Things Are Found**   B2

Chapter 2   **Living Things Depend on One Another**   B38

Unit Project   **Ecosystem Mobile**

Collect nonliving objects from an ecosystem. Do not take any item that you can see is being used by an animal or plant. Use an index card to catalog each item you collect. Sketch and describe the item on the card. Display the objects on a mobile. Use the mobile and cards to tell about interactions within the ecosystem.

Chapter 1

Where Living Things Are Found

LESSON 1
What Are
Ecosystems? B4

LESSON 2
What Are Forest
Ecosystems? B10

LESSON 3
What Is a Desert
Ecosystem? B18

LESSON 4
What Are Water
Ecosystems? B24

Science and
Technology B32

People in
Science B34

Activities for Home
or School B35

CHAPTER REVIEW
and TEST
PREPARATION B36

Living things are all around us—in the air, on the land, and in the water. In this chapter you'll explore where living things are found and how their bodies and behaviors help them live in their environments.

Vocabulary Preview

environment	deciduous forest
ecosystem	tropical rain forest
population	coastal forest
community	coniferous forest
habitat	desert
forest	fresh water
salt water	

FAST FACT

One of the most unusual wildflowers of the southwest United States is the ground cone. It looks like a pine cone sitting on the ground. Since most animals don't eat pine cones, this is a great way of hiding!

FAST FACT

There are about 2,500 black rhinoceroses left in the wild. Because of their small numbers, they are endangered.

Endangered Animals	
Animal	Number Alive
Giant Pandas	1,000
Manatees	1,900
Black Rhinoceroses	2,500
Cheetahs	11,000

LESSON 1

What Are Ecosystems?

In this lesson, you can . . .

INVESTIGATE what makes up an environment.

LEARN ABOUT parts of an environment.

LINK to math, writing, literature, and technology.

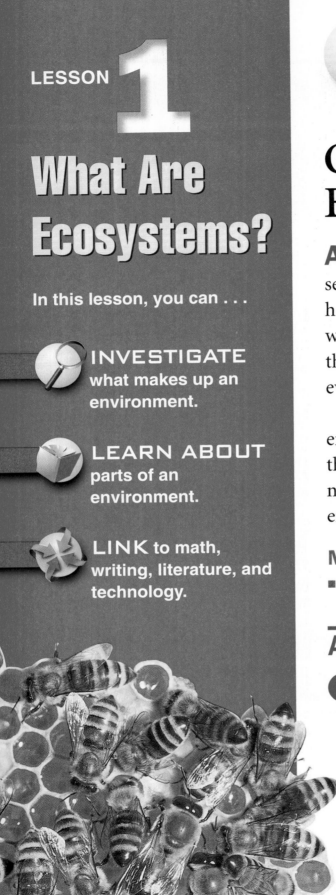

Observing an Environment

Activity Purpose You may have seen trees, insects, and mammals near your home. You may also have seen grass, flowers, worms, and birds. All these things get what they need to live from their *environment*, or everything around them.

In this investigation you will **observe** an environment to find out what kinds of things live there. You will also observe the nonliving things that are part of the environment.

Materials
- wire clothes hanger

Activity Procedure

1. Bend your hanger to make a square. Go outside and place the hanger on the ground. Inside this square is the environment you will **observe**. (Picture A)

◀ These bees live and raise their young in the hive. They make honey to feed the young.

2. Make a list of all the things you **observe**. Next to each thing on your list, **record** whether it is living or nonliving. Write *L* for living and *N* for nonliving.

3. Ask a classmate to share his or her list with you. **Compare** the environments each of you observed.

4. Choose a living thing you **observed** in your hanger environment. Talk with a classmate about which things in the environment help the living thing survive.

Picture A

Draw Conclusions

1. Describe the environment you **observed**.
2. How can you use **observation** to find out how an animal lives?
3. **Scientists at Work** Scientists learn by **observing** and by **gathering data**. They also learn from the data gathered by others. What did you learn about an environment from your classmate's data?

Investigate Further **Observe** a sample of soil in which an earthworm lives. Describe the environment.

Process Skill Tip

When you **observe** living things in their environments, you can **gather data** about the organisms and the environment. The data can help you learn about the environment and about the things that live there.

LEARN ABOUT

Living Things and Their Environments

Where Things Live

Think about where different kinds of plants and animals live. Some kinds of plants and animals live in hot, dry places. Others live in cool, shady forests. Fish live in water. Some insects live in the ground. Others live on plants. In the investigation you observed an environment. An **environment** (en•VY•ruhn•muhnt) is everything around a living thing.

Living things get what they need to live from their environments. They need things such as air, water, food, and shelter. When many kinds of living things share an environment, they must also share the things they need. For example, the plants and animals in a pond all use the air, water, and food from that pond.

✓ **What do living things get from an environment?**

FIND OUT

- how an environment affects a living thing
- what makes up an ecosystem

VOCABULARY

environment
ecosystem
population
community
habitat

A raccoon in a city may find food in trash cans. Many animals can live in more than one environment. ▼

A raccoon in a forest may catch fish to eat. ▶

The Parts of an Ecosystem

In an environment the living and nonliving things that affect each other, or interact, form an ecosystem (EK•oh•sis•tuhm). An ecosystem, such as the pond shown on this page, has many parts.

Frogs live in the pond. All of the frogs, as a group, are called a population. A population (pahp•yoo•LAY•shuhn) is a group of the same kind of living things that live in the same place at the same time.

Populations of insects, waterlilies, and cattails live in the pond, too. Together these populations form a community. A community (kuh•MYOO•nuh•tee) is all the populations that live in an ecosystem.

Within the pond ecosystem, each population lives in a habitat (HAB•ih•tat). A habitat provides a population with all its needs and includes nonliving things and living things. There are many habitats in an ecosystem. For example, the cattails live on the edge of the pond. That is their habitat.

✓ **What makes up an ecosystem?**

The frog population, the cattail population, and all the other populations in this pond form a community.

The pond is the habitat, or home, of all the living things that make up the community.

Nonliving things, such as air, sunlight, water, rocks, and soil, are also part of an ecosystem. The community of living things needs the nonliving things to survive.

The ecosystem of this pond is made up of all the living and nonliving things that interact in and around the pond.

◀ Fires in 1988 destroyed the large trees of this ecosystem. Before 1988 these large trees in Yellowstone National Park were the habitat of many living things.

▲ Ten years after the fire, these wildflowers show that life has returned to Yellowstone.

How Ecosystems Change

Ecosystems can change. Some changes to ecosystems are caused by nature. For example, floods or fires can kill many living things and destroy habitats. But some living things survive when an ecosystem is damaged. A fire doesn't destroy all the seeds in the ground. The seeds that survive grow into new plants. As the plants grow, animals that feed on them return to the area.

A flood can also harm an ecosystem. But a flood brings a new layer of rich soil to a riverbank. New plants grow in the soil.

Ecosystems are also changed by living things. Some changes help the living things meet their needs. For example, people cut down trees and use the wood to build homes and make other products. Beavers cut down trees to build dams and lodges in streams. The other animals that lived in the trees must find new homes.

Many spiders spin webs to use as homes and to help capture food. ▼

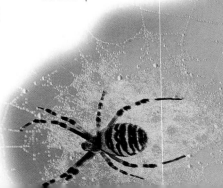

When an ecosystem changes, the animals in it sometimes move to a nearby part of the ecosystem. But sometimes the ecosystem is changed so much that the plants and animals can no longer live there.

✓ **Why do ecosystems change?**

Summary

An environment is everything that surrounds a living thing. The living and nonliving things that interact in an environment make up an ecosystem. An ecosystem is made up of smaller parts that include populations, communities, and habitats, as well as nonliving things, such as air, sunlight, and soil. Nature and living things can change ecosystems.

Review

1. What is an environment?
2. What nonliving things are found in an ecosystem?
3. What is a population?
4. **Critical Thinking** How can building a new road change an ecosystem?
5. **Test Prep** A pond that contains waterlilies, frogs, fish, insects, and cattails is an example of —
   A a population   C a habitat
   B an ecosystem   D a community

LINKS

MATH LINK

Summing Populations
Suppose an ecosystem has 54 squirrels, 12 rabbits, 15 trees, and 100 grasshoppers. How many animals are in the community? How many populations are there?

WRITING LINK

Informative Writing—Report
What if a flood destroyed part of an ecosystem near your home? Write a newspaper report for your class telling what happened to an animal because of the flood.

LITERATURE LINK

Read About Fire To learn more about how fire can affect an ecosystem, read *Wildfires* by Ann Armbruster.

TECHNOLOGY LINK

Learn more about how changes in an ecosystem can affect the living things there by watching *Tainted Water* on the **Harcourt Science Newsroom Video.**

LESSON 2

What Are Forest Ecosystems?

In this lesson, you can . . .

 INVESTIGATE trees in forests.

 LEARN ABOUT different types of forests.

 LINK to math, writing, art, and technology.

Variety in Forests

Activity Purpose Forests have trees. But did you know that there are several types of forests? In this investigation you will **use numbers** to **interpret data** about trees in two kinds of forests.

Materials
- tray of beans, labeled *Tray 1*
- tray of beans, labeled *Tray 2*
- 2 paper cups

Activity Procedure

1. Make a data table like the one shown. Tray 1 stands for the trees in a tropical rain forest. Tray 2 stands for the trees in a deciduous forest. Each kind of bean stands for a different kind of tree.

2. Scoop a cupful of beans from each tray. Carefully pour each cup of beans into its own pile.

◀ Moose like this one live in coniferous forests.

Tropical Rain Forest (Tray 1)		Deciduous Forest (Tray 2)	
Kind of Bean	Number of Beans	Kind of Bean	Number of Beans

3. Work with a partner. One partner should work with the beans from Tray 1. The other partner should work with the beans from Tray 2. (Picture A)

4. Sort the beans into groups so that each group contains only one kind of bean.

5. **Record** a description of each type of bean in the data table. Count the number of beans in each small pile. Record these numbers in the data table.

Picture A

Draw Conclusions

1. How many kinds of "trees" were in each "forest"?

2. Which forest had the most trees of one kind? Why do you think this was so?

3. **Scientists at Work** Scientists learn by **gathering, recording,** and **interpreting data.** What did you learn from your data about variety in forests?

Investigate Further Suppose you want to find out what kinds of trees are most common in your community. Explain how you could **use numbers** to find out.

Process Skill Tip

Using numbers is one way to **record data** from an investigation. When you **interpret data,** you **draw a conclusion** based on the data you have collected.

LEARN ABOUT

Forest Ecosystems

FIND OUT

- about four kinds of forests
- about living things in different kinds of forests

VOCABULARY

forest
deciduous forest
tropical rain forest
coastal forest
coniferous forest

Types of Forests

You probably know that a **forest** is an area in which the main plants are trees. But many other kinds of plants and animals also live in forests.

Forests grow in many parts of the world. Some forests are named for the types of trees that grow in them. Other forests are named for the area in which they grow.

Each type of forest needs a certain amount of rainfall and sunshine. Each also has certain temperatures that let it grow best. If any of these things change, the kinds of plants that grow in the forest may also change.

✓ **What is a forest?**

This graph shows how much rain falls each year in each kind of forest. ▼

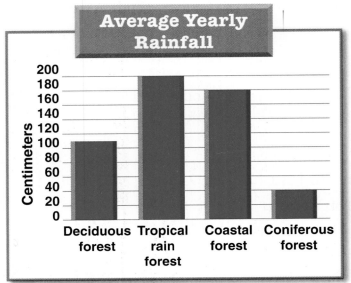

This graph shows the average yearly temperature in each kind of forest. ▼

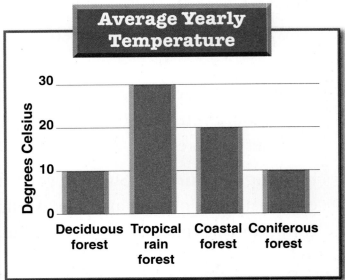

Deciduous Forests

Some trees, such as maples and oaks, have large, flat leaves that drop off each fall. New leaves grow back in the spring. Trees that lose and regrow their leaves each year are called deciduous (dee•SIJ•oo•uhs) trees. Forests made up mostly of these trees are **deciduous forests**. They grow in places that have warm, wet summers and cold winters.

Deciduous leaves change color before they drop in the fall. When a deciduous tree drops its leaves, it needs less water. This helps the tree live through the winter, when water may be frozen.

The deciduous forest is a habitat for many kinds of living things, such as ferns, shrubs, and mosses. Animals such as insects, spiders, snakes, frogs, birds, rabbits, deer, and bears also live here.

✓ **Where do deciduous forests grow?**

▲ During the winter deciduous trees have no leaves.

Deciduous trees grow new leaves in the spring. ▼

In the summer a deciduous forest looks healthy and green. ▼

The leaves of deciduous trees change color just before they drop from the trees in the fall. ▼

Tropical Rain Forests

Tropical rain forests grow in places such as Hawai'i and Costa Rica. These places are hot and wet all year. The trees grow very tall, and their leaves stay green all year.

More types of living things live in rain forests than anywhere else on Earth. Plants and animals make their homes in all the layers of the forest, from the tops of the trees to the ground.

✓ Where do tropical rain forests grow?

◀ This moth lives in the tropical rain forest. It feeds on the plants that grow beneath the trees.

THE INSIDE STORY

Layers of the Rain Forest

A tropical rain forest has three main layers. The top layer is called the *canopy.* It is formed by the branches and leaves of the tallest trees. Below the canopy is the *understory.* The understory is formed by plants that don't grow as tall as trees. The *forest floor* is the lowest layer. Many kinds of plants and animals make their homes in each layer of a tropical rain forest.

Coastal Forests

Coastal forests grow where there is a lot of rain. Unlike a tropical rain forest, a coastal forest grows where it does not get too warm or too cold. But like tropical rain forests, coastal forests are thick with many kinds of tall trees. Coastal forests have the same kinds of layers as tropical rain forests.

✓ **Describe a coastal forest environment.**

▲ The northern spotted owl thrives in the moist, cool environment of the coastal forest.

1. The leaves of the canopy get lots of water and sunlight. Many animals drink water that collects on the leaves.

2. Plants of the understory get less sunlight and water than those in the canopy. Orchids, mosses, and ferns grow on the trunks of the tall trees.

3. Little sunlight reaches the rain-forest floor, and few nutrients are found in the soil. Plants that grow here must find other ways to get nutrients.

Coniferous Forests

What kind of trees would you find where there are very cold winters and cool summers? Mostly, you would find *conifers* (KAHN•uh•ferz)—trees that form seeds in cones. Conifers have needle-like leaves. Pines, spruces, and firs are common conifers. Conifers don't lose their needles in the fall. They stay green all year. This is why conifers are often called *evergreens*. Forests that contain mostly these kinds of trees are **coniferous** (koh•NIF•er•uhs) **forests**.

Conifers grow in areas that get less rain than other types of forests. The needle-shaped leaves of these trees help keep the trees from losing too much water.

Many conifers are shaped like triangles. This shape helps keep heavy snow from piling up on a tree's larger branches in winter, which might cause them to break.

Coniferous forests often have many lakes and streams. The trees, lakes, and streams provide habitats for many animals. Squirrels, moose, and wolves are common. Insects such as mosquitoes and flies also live in coniferous forests.

✓ **What kinds of trees grow in a coniferous forest?**

Coniferous forests are homes for animals such as moose, bears, and wolves. ▼

Summary

There are different types of forest ecosystems. The main types of forests are deciduous forests, tropical rain forests, coastal forests, and coniferous forests. These forests provide habitats for many kinds of plants and animals.

Review

1. What is a forest?
2. Which kind of forest has trees that lose their leaves in the fall?
3. What are the layers of a tropical rain forest?
4. **Critical Thinking** What is the main difference between a tropical rain forest and a coastal forest?
5. **Test Prep** A deciduous forest has —
   A cool summers and warm winters
   B warm, wet summers and dry winters
   C warm, wet summers and cold winters
   D warm, dry summers and dry winters

LINKS

MATH LINK

Interpret a Graph Look back at the first graph on page B12. What is the difference in yearly rainfall between the tropical rain forest and the deciduous forest? Which gets more rain each year?

WRITING LINK

Expressive Writing—Friendly Letter Pick a forest from this lesson. Imagine that you live there. Write a letter to a classmate telling what you like most about your forest.

ART LINK

Leaf Prints Gather leaves from the trees that grow in your area. Dip the leaves into paint, and make leaf prints on paper. Label each print.

TECHNOLOGY LINK

Explore forests as you complete the activity *Backpacking Through Forests* on **Harcourt Science Explorations CD-ROM.**

LESSON 3

What Is a Desert Ecosystem?

In this lesson, you can . . .

 INVESTIGATE a desert ecosystem.

 LEARN ABOUT desert ecosystems.

 LINK to math, writing, health, and technology.

Make a Desert Ecosystem

Activity Purpose A desert is a very dry place. But this ecosystem is home to many kinds of living things. In this investigation you will **make a model** of a desert ecosystem.

Materials
- shoe box
- plastic wrap
- sandy soil
- 2 or 3 desert plants
- small rocks

Activity Procedure

1. **Make a model** of a desert ecosystem. Start by lining the shoe box with plastic wrap. Place sandy soil in the shoe box. Make sure the soil is deep enough for the plants.

2. Place the plants in the soil, and place the rocks around them. Lightly sprinkle the soil with water. (Picture A)

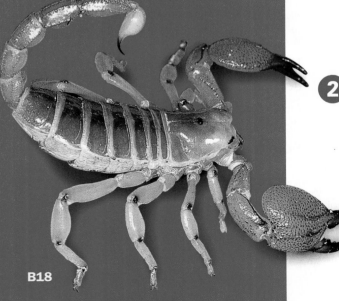

◀ This scorpion is one type of animal that lives in the desert.

3. Place your desert ecosystem in a sunny location. (Picture B)

4. Every two or three days, use your finger to **observe** how dry the soil is. If the soil is *very* dry, add a small amount of water. If the soil is damp, do not add water. Be careful not to water the plants too much.

5. Continue to **observe** and care for your desert ecosystem. **Record** what you observe.

Picture A

Draw Conclusions

1. What kind of environment does your desert ecosystem model?

2. How does **making a model** help you learn about a desert?

3. **Scientists at Work** Scientists often learn by **making models**. What other types of ecosystems can you make models of?

Picture B

Investigate Further How would getting rain every day change a desert ecosystem? **Plan an experiment** in which you could **use a model** to find out.

Process Skill Tip

Observing a desert would be difficult to do in the classroom. To learn more about the real thing, you can **make a model** of a desert.

LEARN ABOUT

Desert Ecosystems

FIND OUT

- about two kinds of deserts
- how plants and animals get what they need in a desert ecosystem

VOCABULARY

desert

Types of Deserts

A **desert** is an ecosystem found where there is very little rainfall. Most deserts get less than 25 centimeters (about 10 in.) of rain each year. Desert plants and animals need very little water to live.

In the investigation you made a model of a hot desert. In summer, hot deserts can have temperatures over 43°C (about 110°F) during the day. At night, the temperatures drop to around 7°C (about 45°F). Most hot deserts have mild winters in which the temperature usually stays above freezing.

Besides the hot deserts you probably know about, there are cold deserts. These deserts have freezing temperatures and blizzards in the winter. But in the hottest months, they are as hot as hot deserts.

✓ **What are the two kinds of deserts?**

The Taklimakan Desert is a cold desert located in western China. ▼

This hot desert is located in the southwestern part of the United States. ▼

THE INSIDE STORY

Parts of a Cactus

The parts of the barrel cactus allow it to live in hot deserts. These parts are common to many other desert plants.

1 A barrel cactus has a thick stem that stores water taken in by the roots. The water is stored in the center part of the stem.

2 Like most cacti, the barrel cactus has a thick skin that is covered with spines. The skin and spines help keep the plant from losing water. The spines also protect the plant from being eaten by animals.

3 The barrel cactus has shallow roots that spread out near the soil's surface. When it rains, these roots quickly soak up the water.

Desert Plants

Deserts are dry places. Plants that grow in deserts have parts that help them save water. Many desert plants grow low to the ground, where it is coolest. Most desert plants have long, shallow roots that spread out near the top of the soil. Here the roots can easily soak up water when it does rain.

Many desert plants have thick stems that store water. Some animals eat these stems to get the water they need to live.

Some desert plants, such as the barrel cactus, have spiny leaves. Spiny leaves help keep the plant from losing water.

✔ **How do desert plants get water?**

Desert Animals

Many animals live in the desert. Desert animals get most of their water by eating plants that store water or by eating other animals.

Reptiles such as snakes and lizards live well in the desert. They stay in the shade during the hot days. During the cold nights they keep warm near rocks, which the sun made hot during the day.

Small mammals such as bats, rabbits, and squirrels also live in deserts. These animals are active at night, when it is cooler. During the day they sleep in their shelters, which shade them from the sun. Some animals burrow into the soil to stay out of the sun.

Scorpions and insects are other desert animals. These animals have hard body coverings that keep them from losing much water.

✓ **How do desert animals get water?**

This sidewinder has dry, scaly skin that helps keep water inside its body. ▼

The shape of this bird's beak helps it get water from the stems of the saguaro cactus. It also uses the saguaro as a home. ▼

◄ Saguaro cactus

Summary

A desert ecosystem has a dry environment. Some deserts are hot, while other deserts are cold. Plants and animals that live in deserts have parts that help them get what they need to live.

Review

1. What is a desert?
2. Where do plants of the hot desert store water?
3. Other than by eating plants, how do desert animals get water?
4. **Critical Thinking** More kinds of animals and plants live in hot deserts than in cold deserts. Why do you think this is so?
5. **Test Prep** The average amount of rain a desert gets in one year is about —
   A 25 mm
   B 25 cm
   C 250 cm
   D 25 m

The large ears of this jackrabbit help it hear enemies. The jackrabbit also gets rid of extra heat from its body through the thin skin on its ears. ▶

MATH LINK

Water from Plants Desert plants store water. To get an idea of how much water a plant can store, squeeze a piece of fruit over a bowl. Measure the amount of juice that comes out.

WRITING LINK

Informative Writing—Compare and Contrast Find out about one desert plant. Write a paragraph for your teacher to compare it to a plant that grows in another ecosystem.

HEALTH LINK

Water and Health Your body needs plenty of water when it is hot outside. With a partner, make a list of 10 things you can do to increase the amount of water you give your body on a hot day.

TECHNOLOGY LINK

Learn more about desert plants and animals by visiting this Internet Site.
www.scilinks.org/harcourt

LESSON 4

What Are Water Ecosystems?

In this lesson, you can . . .

 INVESTIGATE freshwater ecosystems.

 LEARN ABOUT two types of water ecosystems.

 LINK to math, writing, social studies, and technology.

Make a Freshwater Ecosystem

Activity Purpose If you have ever visited a lake, you may know that many kinds of plants and animals live there. Lakes and ponds are freshwater ecosystems. In this investigation you will **make a model** of a freshwater ecosystem.

Materials

- aquarium or other large, clear plastic container
- gravel
- sand
- sheet of paper
- fresh water
- freshwater plants
- rocks
- fish and snails

Activity Procedure

1. Put a layer of gravel at the bottom of the tank. Add a layer of sand on top of the gravel.

◀ Dolphins live, swim, and play in saltwater ecosystems.

B24

2 Set the aquarium in a place where it isn't too sunny. Place a sheet of paper over the sand. Slowly add the water to the tank. Make sure you pour the water onto the paper so the sand will stay in place. (Picture A)

3 Remove the paper, and put the plants and rocks into the tank. Let the tank sit for about one week. After one week, add the fish and snails. (Picture B)

4 **Observe** and care for your freshwater ecosystem.

Picture A

Picture B

Draw Conclusions

1. What are some things you **observed** in your freshwater ecosystem?

2. Why do you think you waited to add the fish to the tank?

3. **Scientists at Work** Scientists often **make a model** of an ecosystem so they can **observe** it in a laboratory. How did making a model help you observe a freshwater ecosystem? How is your model different from a real pond?

Investigate Further What other kinds of plants and animals might live in a freshwater ecosystem? To find out, visit a pet shop that sells fish. Make a list of the freshwater plants and animals you find there.

Process Skill Tip

It can be difficult to **observe** all the living things in a freshwater pond or lake. You can learn about the real thing by **making a model** of a freshwater ecosystem.

B25

Water Ecosystems

Types of Water Ecosystems

FIND OUT

- about freshwater and saltwater ecosystems
- what plants and animals live in water ecosystems

VOCABULARY

salt water
fresh water

Water covers more than 70 percent of Earth's surface. When you look at a globe, you can see that most of this water is in oceans and seas. Oceans and seas contain **salt water**, or water that has a lot of salt in it. Marshes and a few lakes also have salt water in them. These ecosystems are called *saltwater ecosystems*.

In the investigation you made a model of a freshwater ecosystem. **Fresh water** is water that has very little salt in it. Lakes, rivers, ponds, streams, and some marshes are *freshwater ecosystems*.

✔ What are two types of water ecosystems?

The blue parts of this map show the major water ecosystems of the world. ▼

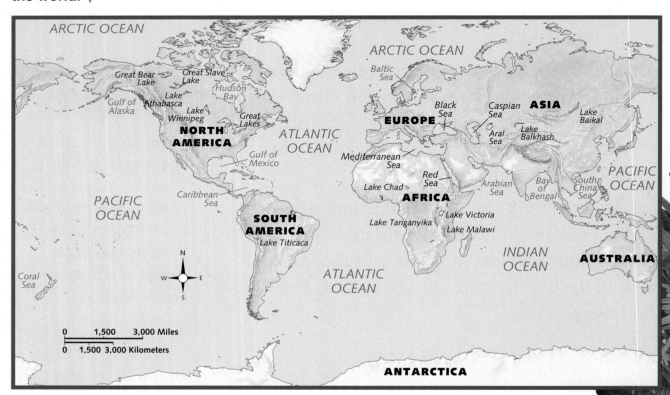

Saltwater Ecosystems

There are many different saltwater ecosystems because salt water is not the same everywhere. The amount of salt in ocean water is different in different places. Near the surface, rain can make the water less salty. Water is also less salty near shores where fresh water from rivers and streams flows into oceans. But no matter where you are in the ocean, the water has more salt in it than fresh water has.

Ocean water that is near shore or at the surface is often warmer than water deep in the ocean. This is because the sun warms the water in these areas. At the shoreline, salt water may collect in tide pools. Many plants and animals live in this warm, shallow water.

Barnacles (BAR•nuh•kulz) are small animals that live out their lives stuck to rocks and other objects. ▼

✓ **Why might salt water be less salty near shore?**

◀ A tide pool is a saltwater ecosystem that forms at a shoreline. ▼

◀ Some anemones (uh•NEM•uh•neez) live in the warm, shallow waters of tide pools.

Ocean Ecosystems

The oceans are Earth's largest ecosystems. Because they are so large, many kinds of plants and animals live in them. All these living things are suited to life in salt water.

Not all parts of the ocean are the same. The saltiness of the water can be different. Water is deeper in some places than in others. The water temperature also changes from place to place. All these differences, along with how deep into the water the sunlight can reach, affect the kinds of living things that are found in different parts of the ocean.

✓ **What do all ocean plants and animals have in common?**

THE INSIDE STORY

Ocean Zones

The ocean can be divided into zones. Each zone is defined by how much sunlight it gets. Areas that get the most sunlight usually have the warmest temperatures and more plant and animal life. Areas that receive little sunlight are dark and have few animals and very few plants. They also have very cold water.

This underwater kelp forest grows in shallow water. It provides food and shelter for many different fish and other ocean animals.

◀ These tropical fish live in areas of warm water.

Where water is shallow, sunlight reaches the ocean bottom. The sunlight allows many plants to grow in this zone. Animals that feed on these plants also live here.

Farther from shore, sunlight cannot reach the ocean bottom. Plants in this part of the ocean float near the surface, where they can get sunlight. Many swimming animals come to the surface to feed. Others feed on the animals that eat the plants.

This angler fish lives in the deep part of the ocean. ▼

Sunlight cannot reach the floor of the deep ocean. This part of the ocean has almost no plants. Water is very cold here. Because there are very few plants (if there are any at all), the deep ocean also has the fewest animals.

Freshwater Ecosystems

Not all freshwater ecosystems are the same. The main kinds of freshwater ecosystems include rivers, streams, lakes, and ponds.

Many plants and animals live in or around rivers and streams. Rivers and streams have moving fresh water. The water may move quickly or slowly. How fast the water moves helps determine what living things can survive in the water.

The water in lakes and ponds is still. Lakes are usually larger and have deeper water than ponds. Water temperature depends on where the lake or pond is and how deep the water is.

As in an ocean, most plants and animals in a lake or pond live in the shallow water. The fewest plants and animals live where the water is too deep for sunlight to reach the bottom.

▲ This cooter turtle is a common shore animal of ponds, rivers, marshes, and lakes in Florida.

✓ **What are two kinds of freshwater ecosystems?**

The American egret is a wading bird that lives near freshwater ecosystems such as lakes. ▼

A young mayfly can stick to a rock or a plant so fast-moving water won't wash it away. ▼

The bullhead catfish makes its home in a lake. ▼

Summary

Water ecosystems may have salt water or fresh water. An ocean is a saltwater ecosystem. Rivers, streams, lakes, and ponds are freshwater ecosystems. Different kinds of plants and animals live in and around these water ecosystems.

Review

1. What are the two kinds of water ecosystems? Name two ways they are different.
2. In a pond, where do most of the animals live?
3. Which zone of the ocean gets the least amount of sunlight?
4. **Critical Thinking** What would happen to the animals and plants in a saltwater ecosystem if it were flooded by fresh water? Explain your answer.
5. **Test Prep** Which is **NOT** a freshwater ecosystem?
   - **A** river
   - **B** lake
   - **C** pond
   - **D** tide pool

MATH LINK

Using Graphs Use a computer graphing program such as **Graph Links** to make a bar graph of the following water depths: river, 1 meter; lake, 6 meters; ocean, 122 meters. How much deeper is the ocean than the lake?

WRITING LINK

Informative Writing— Narration Suppose you are a tour guide on a submarine in the deep ocean. Write a narration that describes to your passengers the kinds of plants and animals you would expect to find living there.

SOCIAL STUDIES LINK

Mapping Water Ecosystems Make an outline map of your state. Draw the major bodies of water. Label them as fresh or salt water.

TECHNOLOGY LINK

Visit the Harcourt Learning Site for related links, activities, and resources.
www.harcourtschool.com

SCIENCE AND TECHNOLOGY

Using Computers to Describe the Environment

Do you know what kind of environment you live in? Students all over the world are using computers to help them learn more about the plants, the animals, and other things in their environments.

Ecosystem Movies

Students at Gilliland Elementary School in Blue Mound, Texas, have been using computers to create a multimedia field guide for their hometown. They take careful notes during field trips into the Blackland Prairie near their school. They take photographs and make audio and video recordings. Afterward, the students combine everything on a computer to make the field guide.

People who use the field guide can see photographs of the wildflowers in the area. They can also watch videos of the class and hear students talk about the history of the prairie.

A World of Information

Students at many schools belong to a program called GLOBE, or Global Learning and Observations to Benefit the Environment. The GLOBE program includes students, teachers, and scientists from all over the world.

First, GLOBE scientists teach teachers how to take samples and make measurements. Then the teachers teach their students how to do it. The teachers and students regularly collect data from their own environments. They measure and record information about soil, water, air, and plants in their area. They enter all their data onto the main GLOBE Web site.

Mapmaking on the Internet

The scientists at GLOBE put together the data from all of the schools. They use the data to learn about specific areas and about patterns occurring all over the world. Because so many schools are sending in data, the scientists can learn much more than they would if they were working on their own. One of the things they do with the data is make environmental maps of the world. Students and teachers can log onto the GLOBE Web site to see these maps.

Think About It

1. What would you include in a field guide for your environment?
2. What kinds of maps could GLOBE scientists make from students' information?

WEB LINK:
For Science and Technology updates, visit the Harcourt Internet site.
www.harcourtschool.com

Careers — Teacher

What They Do Teachers work with children or adults to help them learn. They often use computers and other tools to make learning new ideas fun.

Education and Training Almost all teachers have college degrees. Most colleges offer courses that help students learn how to teach. Teachers specialize in the subject or the grade level they want to teach.

PEOPLE IN SCIENCE

Margaret Morse Nice
ORNITHOLOGIST

"It was an unknown world and each day I made fresh discoveries."

Ornithologists are people who study birds. Perhaps, like Margaret Morse Nice, you have watched birds in your back yard. By the age of 12, she was recording her observations about birds. What began as fun became her lifework.

After college, Morse married Leonard Nice, whom she had met there. Each time the family moved, Margaret Nice studied the birds around her home.

Nice was one of the first to use colored bands to identify birds. To follow birds over time, she would first capture a bird. Then she would place a colored band of plastic on the bird's leg and let the bird go. In this way, Nice studied how birds mated, built nests, and raised their young.

Nice and her husband wrote a book together on the birds of Oklahoma. Nice later published a two-part study of song sparrows. She wrote more than 250 articles for magazines during her lifetime. She also translated into English many of the articles that people from other countries had written about birds.

Think About It

1. Why is it helpful to have someone translate articles written by scientists in other countries?
2. What birds can you observe from your home?

ACTIVITIES FOR HOME OR SCHOOL

Earthworm Habitat

What is the habitat of an earthworm?

Materials
- clear plastic container
- garden soil (not potting soil)
- 2 to 3 earthworms
- small rocks, sticks, and leaves
- wax paper
- water

Procedure
1. Loosely spread the soil on the wax paper. Add the rocks, sticks, and leaves. Add a small amount of water so the soil is moist but not wet. Mix these materials together.
2. Add the soil mixture to the plastic container. Carefully add the earthworms to the loose soil. Place your earthworm habitat in a warm, dark place. Keep the soil moist.

Draw Conclusions
After observing your earthworm habitat for a week, make a list of the things you think the habitat you provided gave your earthworms that they needed to live.

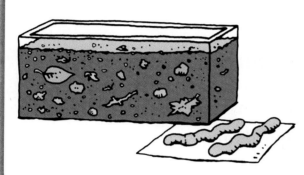

Salt Water and Fresh Water

How are salt water and fresh water different?

Materials
- small jar
- water
- egg in the shell
- small spoon
- salt

Procedure
1. Half fill the jar with water. Put the egg in the jar. Record what happens to the egg.
2. Remove the egg, and stir a spoonful of salt into the water. Put the egg back into the water, and record what happens to the egg.
3. Continue adding salt to the water until you observe a change.

Draw Conclusions
How do you think salt in the water affects the animals living in the ocean?

Chapter 1 Review and Test Preparation

Vocabulary Review

Use the terms below to complete the sentences 1 through 12. The page numbers in ( ) tell you where to look in the chapter if you need help.

environment (B6)
ecosystem (B7)
population (B7)
community (B7)
habitat (B7)
forest (B12)
deciduous forest (B13)
tropical rain forest (B14)
coastal forest (B15)
coniferous forest (B16)
desert (B20)
salt water (B26)
fresh water (B26)

1. An area where the main plants are trees is called a ___.
2. All the living things of the same kind that live in the same area at the same time make up a ___.
3. The populations that live together in the same place make up a ___.
4. An ecosystem that is very dry is a ___.
5. The interactions between the living and nonliving parts of the environment make up an ___.
6. To a population of frogs living near a pond, the pond is its ___.
7. Rivers and most lakes are ___ ecosystems.
8. A ___ has warm summers and cool winters. The trees drop their leaves in the fall.
9. Two types of forests that get a lot of rain are a ___ and a ___.
10. An ___ is everything that surrounds a living thing.
11. The winters are cold in a ___, and the trees have needle-shaped leaves.
12. The water in the ocean is ___.

Connect Concepts

Write the terms from the Word Bank where they belong in the concept map.

tropical rain   coastal   forests
coniferous   deciduous

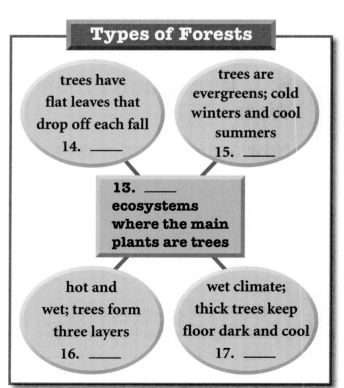

Types of Forests

trees have flat leaves that drop off each fall
14. ___

trees are evergreens; cold winters and cool summers
15. ___

13. ___ ecosystems where the main plants are trees

hot and wet; trees form three layers
16. ___

wet climate; thick trees keep floor dark and cool
17. ___

Check Understanding

Write the letter of the best choice.

18. What might happen to a forest that is damaged by fire?
    - A It will grow back.
    - B It is destroyed forever.
    - C It will stay the same.
    - D It will fill up with water and become a marsh.

19. Which characteristic of conifers helps keep them from losing water?
    - F broad, flat leaves
    - G thick, fleshy leaves
    - H triangular shape
    - J needle-shaped leaves

20. What are the two types of desert?
    - A dry and wet
    - B hot and cold
    - C fresh and salt
    - D tropical and coastal

21. What are the two types of water ecosystems?
    - F fresh and salt
    - G hot and cold
    - H hot and dry
    - J deciduous and coniferous

Critical Thinking

22. Why might a raccoon hunt for food in a trash can instead of in a forest?

23. How can the building of new homes and roads by humans affect the plants and animals in an area?

Process Skills Review

24. In Lesson 1 you used your **observations** to **gather data** about organisms in their environment. What are some of the ways that you gathered data?

25. Find three examples in the chapter where you could **use numbers** to **record data**. Why is this a good way to **interpret data**?

26. In Lessons 3 and 4 you **made a model** of two ecosystems—a desert and a pond. How did either model help you learn about the features of these ecosystems?

Performance Assessment

Diagram a Rain Forest

Work with two or three other students. On a large sheet of paper, draw a rain-forest ecosystem. Show the three layers, and label each. Then draw or describe two or more animals that live in each layer.

Chapter 2
Living Things Depend on One Another

LESSON 1
How Do Animals Get Food? **B40**

LESSON 2
What Are Food Chains? **B46**

LESSON 3
What Are Food Webs? **B52**

Science Through Time **B58**

People in Science **B60**

Activities for Home or School **B61**

CHAPTER REVIEW and TEST PREPARATION **B62**

Vocabulary Preview

interact
producer
consumer
decomposer
food chain
energy pyramid
food web
predator
prey

Like all living things, you depend on plants and animals around you to meet your needs. You eat plants and animal products. You may live in a building made from wood. You wear clothes made from plants fibers. Plants and animals depend on one another to help them meet their needs.

FAST FACT

Some areas of the world have more living things than others. If you want to get an idea of how productive an area is, you can compare how many plants live there with how many live in other areas.

Plant Material in Different Environments

Area	Plant matter per square meter
Tropical Rain Forest	2000 g
Grassland	800 g
Arctic Tundra	140 g
Desert	80 g

Tropical rain forest

FAST FACT

Hippopotamuses live in the rivers of central Africa. They eat grasses and other plants that grow in the water. The wastes they produce are rich in nutrients, and they help the plants to grow.

LESSON 1

How Do Animals Get Food?

In this lesson, you can...

INVESTIGATE how animals use their teeth to help them get food.

LEARN ABOUT how living things get food.

LINK to math, writing, health, and technology.

Animal Teeth

Activity Purpose Bite into an apple. Which teeth do you use? Which teeth do you use to chew the apple? In this investigation you will **observe** the shapes of teeth of different animals.

Materials
- blank index cards
- books about animals

Activity Procedure

1 Observe the pictures of the animals. Look closely at the shape of each animal's teeth.

2 Use one index card for each animal. **Record** the animal's name, and draw the shape of its teeth.

◀ Animals that are pets get their food from people. But animals that are wild must find their own food.

Shark ▼

Bobcat ▼

3. With a partner, make a list of words that describe the teeth. **Record** these words next to the drawings on the index cards. (Picture A)

4. Think about the things each animal eats. Use books about animals if you need help. On the back of each index card, make a list of the things the animal eats.

Picture A

Draw Conclusions

1. Which animals might use their teeth to catch other animals? Which animals might use their teeth to eat plants? Explain.

2. Some animals use their teeth to help them do other things, too. **Observe** the beaver's teeth. How do its teeth help it cut down trees?

3. **Scientists at Work** Scientists learn by **observing**. Scientists can learn how animals use their teeth by watching how and what the animals eat. From what you observed in this investigation, what can you **infer** about the shapes of animals' teeth?

Process Skill Tip

Observing and inferring are not the same. When you **observe**, you use your senses. When you **infer**, you form an opinion using your observations.

Wolf ▼

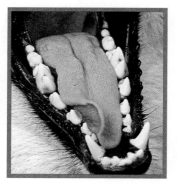

Horse ▼

Beaver ▼

LEARN ABOUT

Living Things and Food

Making and Getting Food

FIND OUT

- how plants and animals interact with the environment
- how living things get food

VOCABULARY

interact
producer
consumer
decomposer

All living things need food. In the investigation you saw that an animal's teeth match the food it eats. But not all living things have teeth. Some living things get their food in other ways. For example, a bird uses its beak to get food. Plants make their own food.

Plants and animals work together, or **interact** (in•ter•AKT), with the environment to get what they need. Plants interact with sunlight, air, and water to make food. Animals interact with plants or other animals to get their food. Animals also interact with nonliving things in their environment, such as water, sunlight, soil, and rocks. For example, a snake lies in the sun for warmth and drinks water. The same snake may make its home in the soil or under a rock.

✓ **Why do plants and animals interact with the environment?**

This strawberry plant uses energy from the sun to make its own food. ▼

This strawberry provides a chipmunk with some of the energy it needs to live. ▼

The chipmunk provides energy for a snake that catches and eats it. ▼

Producers

Plants are producers. A **producer** (proh•DOOS•er) is a living thing that makes its own food. Producers use the food they make to live and to grow.

Plants make more food than they need. This extra food is stored in roots, leaves, seeds, and fruit. People and other animals then eat this stored food as their own food.

✓ **What is a producer?**

▲ These foods come from plants. Plants are producers that make their own food. Animals then eat plants as food.

Consumers

Animals cannot make their own food. They must eat plants or other animals. An animal is a **consumer** (kuhn•SOOM•er). A consumer is a living thing that eats other living things as food.

Consumers can be grouped by the kinds of food they eat. Some consumers eat only plants. Sheep, like this bighorn sheep, eat the leaves, twigs, fruits, and nuts of many plants.

A few consumers eat only one kind of plant. Giant pandas eat only bamboo. To survive, they must live where bamboo grows.

Animals that eat only plants may have body parts that help them eat. For example, the giant panda has an extra bone in its hand that helps it hold bamboo as it eats.

The bighorn sheep is a consumer that eats grass. ▼

Some animals get food by eating other animals. Often they must hunt and kill their food. Animals that get food in this way have body parts that help them catch and eat their food. For example, an owl has strong claws that it uses to catch animals. Its sharp beak helps it tear meat.

Some animals eat both plants and other animals. A box turtle eats both berries and insects. The meats you eat come from animals and the vegetables you eat come from plants.

✓ **What is a consumer?**

This owl catches small animals for its food. ▼

Decomposers

A **decomposer** (dee•kuhm•POHZ•er) is a living thing that breaks down dead things for food. Decomposers also break down the wastes of living things. As decomposers feed, they help clean the environment. Two decomposers you may know of are fungi, such as mushrooms, and earthworms.

✓ **What is a decomposer?**

▲ Many bacteria are decomposers.

The fungi growing on this log are decomposers. The fungi are using the dead log as food. ▼

Summary

Plants and animals interact. They depend on their environments and on one another to get the food they need. Plants are producers. Animals are consumers. Decomposers get food by breaking down wastes or dead things.

Review

1. How do producers get their food?
2. How do consumers get their food?
3. What are the three groups of consumers?
4. **Critical Thinking** How do decomposers help keep the environment clean?
5. **Test Prep** Which of the following is **NOT** a consumer?
   - **A** bird
   - **B** tree
   - **C** squirrel
   - **D** human

LINKS

MATH LINK

Consumers on an Island Scientists studied 60 moose on an island. They found that during the summer, each moose eats the fruit of 25 blackberry bushes. What is the minimum number of blackberry bushes on the island?

WRITING LINK

Informative Writing— Explanation Suppose that the bushes on the island in the Math Link all die. Write a paragraph for your teacher explaining what you think might happen to the moose population.

HEALTH LINK

Teeth Think about how you use your teeth to eat. Which teeth do the cutting? The chewing? How are the teeth different?

TECHNOLOGY LINK

Learn how scientists help injured animals get food by visiting the Smithsonian Institution Internet Site.
www.si.edu/harcourt/science

Smithsonian Institution®

LESSON 2

What Are Food Chains?

In this lesson, you can . . .

INVESTIGATE a food-chain model.

LEARN ABOUT food chains and energy pyramids.

LINK to math, writing, literature, and technology.

Make a Food-Chain Model

Activity Purpose You get energy from the food you eat. All living things get energy from the food they eat. In this investigation you will **make a model** to show how living things interact to get their energy from food.

Materials
- index cards
- marker
- 4 pieces of yarn or string
- tape

Activity Procedure

1. In the bottom right-hand corners, number the index cards 1 through 5.

2. On Card 1, draw and label grass. On Card 2, draw and label a cricket. On Card 3, draw and label a frog. On Card 4, draw and label a snake. On Card 5, draw and label a hawk. (Picture A)

◀ A puffin eats different kinds of sea animals. It can catch as many as ten small fish at one time in its beak.

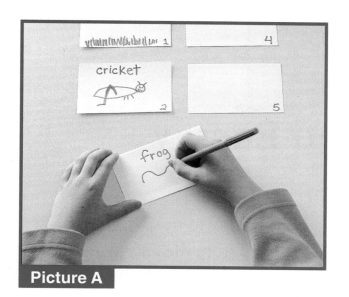
Picture A

Picture B

3. Order the cards in a line with Card 1 first and Card 5 last. Use yarn and tape to connect the cards. (Picture B)

4. Stretch the connected cards out on a table. The cards form a model called a food chain.

5. Discuss with a classmate how each living thing in the food chain gets its food. Tell which things in your model show producers. Tell which things show consumers.

Draw Conclusions

1. In your model, which living thing is last in the food chain? Why do you think it is in this place?

2. In which part of the food chain is the producer found? Why do you think it is there?

3. **Scientists at Work** Scientists **use models** to help them study things in nature. How does using a model of a food chain help you understand living things and the food they eat?

Process Skill Tip

It is hard to **observe** a real-life food chain. **Using a model** of a food chain helps you learn about the real thing.

LEARN ABOUT

FIND OUT
- how living things get energy
- how energy moves through a food chain

VOCABULARY
food chain
energy pyramid

Food and Energy

Food Chains

All living things need energy to live. Producers get energy from sunlight. They store the energy in the food they make. Consumers can't make their own food. They get their food by eating other living things. In this way, the consumers get the energy they need.

In the investigation you saw that the path of food from one living thing to another forms a **food chain**. A food chain also shows how energy moves through the environment. For example, grass uses the energy in sunlight to make its food. A cricket that eats the grass gets the energy stored in the grass. If a frog eats the cricket, it gets energy that is stored in the cricket. In this way, the energy that started with the sun is passed from the grass to the cricket to the frog.

✓ **What is a food chain?**

This turtle eats slugs that eat leaves. ▶

A Food Chain

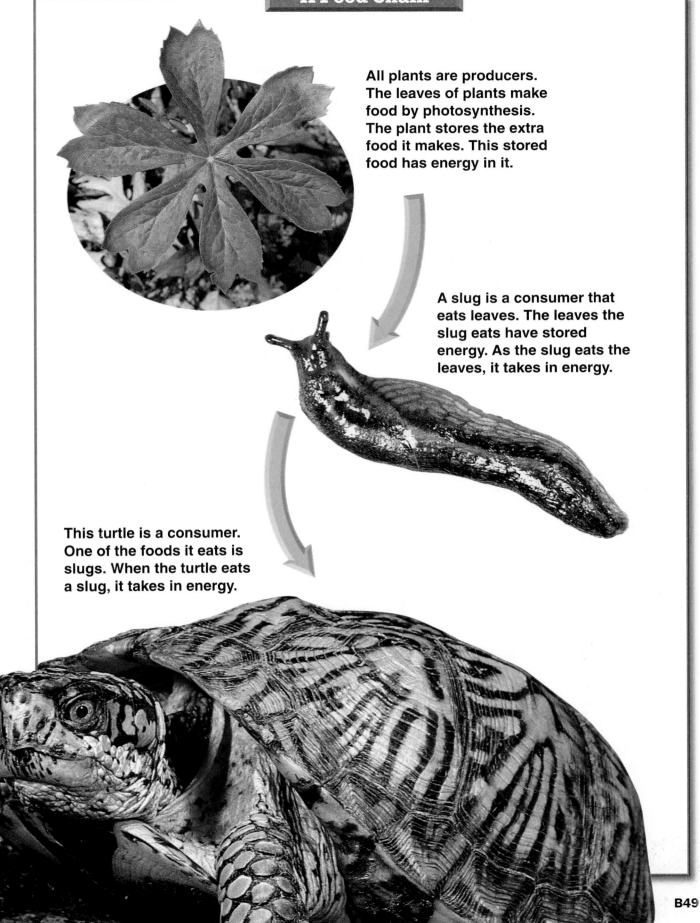

All plants are producers. The leaves of plants make food by photosynthesis. The plant stores the extra food it makes. This stored food has energy in it.

A slug is a consumer that eats leaves. The leaves the slug eats have stored energy. As the slug eats the leaves, it takes in energy.

This turtle is a consumer. One of the foods it eats is slugs. When the turtle eats a slug, it takes in energy.

Energy from Food

Every living thing uses energy to live and to grow. The energy that a living thing uses cannot be passed along through the food chain. Because of this, the higher on the food chain a living thing is, the less energy there is.

An **energy pyramid** shows that the amount of useable energy in an ecosystem is less for each higher animal in the food chain.

In an energy pyramid, there are more producers than any other kind of living thing. Most of the energy in an ecosystem is found in plants. Animals that eat plants make up the next level. The upper parts of the pyramid are made up of animals that eat other animals. The higher in the pyramid an animal is, the fewer of that animal there are. This is because there is less energy available to them.

✓ **What is an energy pyramid?**

1. Energy from the sun is taken in by plants and other producers. This energy is used by plants for growth and to make fruits and seeds. Energy that is not used by the plant is stored.

2. Animals that eat plants are called first-level consumers. These animals must eat many plants to get the energy they need. Energy not used by the animal is stored in its body.

3. Animals that eat other animals are called second-level consumers. There are fewer of these animals. Some of the energy that is stored in first-level consumers can be passed on to these animals.

4. There are very few animals at the top of the pyramid. The energy these top-level consumers get has been passed through all the other parts of the food chain.

THE INSIDE STORY

An Energy Pyramid

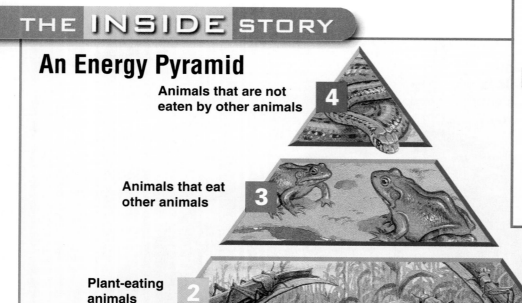

- Animals that are not eaten by other animals — 4
- Animals that eat other animals — 3
- Plant-eating animals — 2
- Plants — 1

▲ A hyena is a *scavenger,* an animal that feeds on dead animals.

Summary

Living things get their energy from food. Animals cannot make their own food, so they eat other living things to get energy. A food chain is the flow of food in an ecosystem from one living thing to another. An energy pyramid shows that the amount of useable energy in an ecosystem is less for each higher animal in the food chain.

Review

1. How does energy get from a producer to a meat-eating consumer?
2. What kind of living thing is at the top of a food chain?
3. Where is the most food energy in an energy pyramid found?
4. **Critical Thinking** What is the source of all the energy on Earth?
5. **Test Prep** Which is passed in a food chain from one living thing to another?

   A producers   C sunlight
   B animals   D energy

LINKS

MATH LINK

Animals in the Food Chain
One food chain is made up of a producer, a consumer that eats plants, a consumer that eats both plants and animals, and a consumer that eats only animals. What fraction of living things in this food chain eats plants?

WRITING LINK

Expressive Writing—Poem
You are part of many food chains. Pick your favorite dinner. Write a poem for your family about how the different foods fit into food chains.

LITERATURE LINK

Underwater Food Chains
Plants and animals that live in water are parts of food chains, too. In the book *Mangrove Wilderness,* Bianca Lavies tells about food chains in an estuary.

TECHNOLOGY LINK

To learn more about food chains, watch the video *Poisoned Eagles* on the **Harcourt Science Newsroom Video.**

LESSON 3

What Are Food Webs?

In this lesson, you can . . .

INVESTIGATE food webs.

LEARN ABOUT how food chains overlap to form food webs.

LINK to math, writing, literature, and technology.

Make a Food Web

Activity Purpose Most animals eat more than one kind of food. Because of this, one living thing can be a part of more than one food chain. Food chains in an ecosystem overlap to form *food webs*. In this investigation you will **make a model** of a food web.

Materials
- index cards, cut into fourths
- poster board
- tape or glue
- crayons

Activity Procedure

1. Write the name of each living thing from the chart on its own card.

2. Glue the cards onto a sheet of poster board so they form a circle. Leave room for writing. (Picture A)

◀ A sea otter is part of an ocean food web. It feeds on abalone and other shellfish. As the otter floats on its back, it cracks open the shellfish by banging it against a rock it carries on its chest.

Living Thing	What It Eats
clover	uses the sun to make its own food
grasshopper	clover
frog	grasshopper
snake	frog, mouse
owl	snake, mouse
mouse	clover

Picture A

3. Look at the chart again. List two different food chains you could make.

4. Draw arrows between the parts of each food chain. Use a different color for each food chain. You have now made a food web. (Picture B)

5. **Observe** your model to see how the food chains overlap. What other living things could you add to your food web?

Picture B

Draw Conclusions

1. What is the producer in this food web?
2. What does your food web tell you about producers and consumers?
3. **Scientists at Work** Scientists sometimes **make models** to help them learn about things. How did drawing a food web help you learn about animals in a real ecosystem?

Investigate Further Cut out magazine pictures of different plants and animals. Work with a partner to make a food web that includes these plants and animals.

Process Skill Tip

It is hard to observe a real-life food web. **Using a model** of a food web helps you learn about the real thing.

B53

LEARN ABOUT

Food Webs

FIND OUT
- about food webs
- how living things interact in food webs

VOCABULARY

food web
predator
prey

Predator and Prey

There are many food chains in an ecosystem. Sometimes these food chains overlap. A model that shows how food chains overlap is called a **food web**. A food web contains producers and consumers that are used as food by more than one living thing.

Food webs contain animals that eat other animals. An animal that hunts another animal for food is called a **predator** (PRED•uh•ter). The animal that is hunted is called **prey** (PRAY). Some animals can be both predator and prey. For example, when a snake eats a mole, the snake is the predator. The mole is the prey. If the snake is eaten by a hawk, the snake becomes the prey. The hawk is the predator.

✓ **What is a food web?**

This marsh ecosystem has many kinds of animals. These animals may be predators, prey, or both. ▼

Marsh Ecosystem

Many overlapping food chains make up the food web for this marsh ecosystem.

1 The plants in the marsh use the energy of the sun to make food.

2 Insects eat the plants that grow in the marsh.

3 Fish eat the insects. Some fish also eat the plants. Smaller fish are eaten by larger fish.

4 Birds eat the fish in the marsh. Some birds also eat the insects and the plants that grow in the marsh.

5 Alligators eat both fish and birds. Alligators are top-level consumers. They are not eaten by other animals in the marsh.

What other food chains can you identify in this food web?

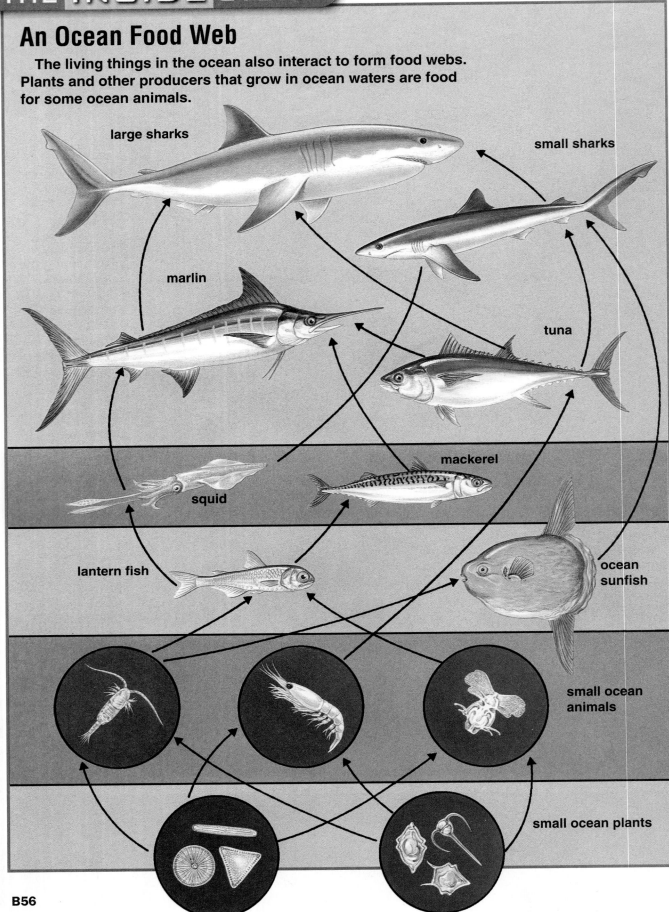

Summary

Most ecosystems have many food chains. These food chains overlap and link together to make food webs. An animal in a food web may be a predator, prey, or both.

Review

1. How are food webs and food chains alike? How are they different?
2. How can an animal be both predator and prey?
3. How does an alligator get energy?
4. **Critical Thinking** In the following food chain, name the predators and the prey.

   insect → frog → snake
5. **Test Prep** Which of the following is a producer in an ocean ecosystem?

   A shark   C starfish
   B seaweed   D fish

Other animals prey on these shrimp. Just like animals that live on land, many ocean predators have body parts that help them catch and eat their prey. ▼

MATH LINK

Recording Food Web Data Choose a food web from this lesson. Count the different food chains in that web. How many animals in the web are predators, and how many are prey? Record your findings in a table.

WRITING LINK

Informative Writing—Explanation Study the ocean food web on page B56. Pick one of the animals and write a paragraph for your classmates that explains how the animal fits into a food chain and a food web.

LITERATURE LINK

Read About Predators Learn about some amazing predators, such as a fishing spider or a vampire bat. Read *Extremely Weird Hunters* by Sarah Lovett.

TECHNOLOGY LINK

Visit the Harcourt Learning Site for related links, activities, and resources.
www.harcourtschool.com

SCIENCE THROUGH TIME

People and Animals—A Long Relationship

Animals and humans have had a long relationship. People learned many thousands of years ago to domesticate, or tame, animals. Animals have also been hunted for food and clothing for thousands of years.

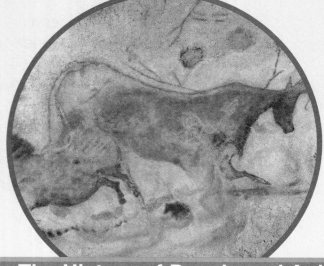

Uses of Animals

Animals are part of the web of life. They are consumers, living off plants and smaller animals. But people use animals and animal products as food. Some people choose not to eat animals. Many of those people do eat products that come from animals, such as milk, eggs, or honey. All of us depend on animals to pollinate the flowers of many fruits and vegetables.

Animal fur or skins may be used for clothing. Perhaps some of the clothing you are wearing today came

The History of People and Animals

15,000 B.C.
Horses are drawn in cave paintings in France.

8000 B.C.
People begin to tame animals.

1400 B.C.
Hittites who live in Turkey train horses.

1519
Spanish explorers bring horses to North America.

Timeline: 15,000 B.C. — 7500 B.C. — B.C./A.D. — 1500 — 1600 — 1700

from an animal. We get wool from sheep. Silkworms make silk.

Over the centuries, people have used animals to work. Large animals like oxen can be used to plow fields. Horses, camels, and elephants are good for carrying people and goods. Some dogs can guard, rescue, hunt, herd, or guide. Cats catch mice.

For some of us, domestic animals are best friends. You may have a cat or a dog as a family pet. Some people have more unusual pets, such as small reptiles or birds from foreign countries.

The Camel—A Very Useful Animal

The camel is one animal that is very useful to humans. Most camels now live in the deserts of Asia and Africa. But scientists have evidence that camels lived in North America before the Ice Age. They died out before Europeans came to the continent. But the U.S. Army brought camels back to North America to carry cargo from Texas to California during the mid-1800s. The railroad was faster than camels, though. After the railroad across the country was finished, most camels went to live in zoos and circuses.

People in Asia still use camels to carry heavy loads, especially in desert areas. One camel can carry more than 136 kilograms (300 lb). A working camel travels about 40 kilometers (25 mi) a day at about 5 kilometers per hour (3 mph). Also, camels can go for a long time without water.

Camels provide meat and milk. People make cheese and butter from their milk. Camel hair makes warm blankets, clothes, and tents.

As you can see, the relationship of animals and humans is complicated. Think of all the types of animals you have studied—fish, reptiles, birds, amphibians, and mammals. Then look around at home and at school. What should you thank an animal for today?

Think About It

- Why do you think people train animals to do specific jobs?

1850
Camels arrive in the United States.

1800　　　1900　　　2000

1973
The United States passes the Endangered Species Act to protect animals.

PEOPLE IN SCIENCE

Akira Akubo
OCEANOGRAPHER

Growing up in Japan, a country surrounded by water, Akira Akubo became interested in the ocean. His interest led him to study oceanography.

Questions Akubo has studied include why and how fish live in schools. Fish gather in schools for protection. A school may break up at night to feed, but the fish gather again the next morning. A school may have as few as two dozen fish or as many as several million. All the fish in a school are about the same size. Adult fish and young fish are never in the same school. Some fish form schools when they are young and stay together all their lives. Other species of fish form schools for only a few weeks after hatching.

Akubo has also studied plankton—tiny animal-like and plantlike living things that float near the water's

surface. Most plankton are so small that they can be seen only with a microscope. Plankton is food for many other living things in the sea. Animal-like plankton eat the plantlike plankton. A lot of plankton is eaten by fish. Some whales eat nothing but tons of plankton!

Akubo is interested in land animals, too. He believes that studying land animals can help him learn more about animals in the water. He hopes that comparing ocean animals with land animals will help him predict animal behavior.

Think About It
1. Why is plankton important?
2. What is the advantage for fish of traveling in schools?

ACTIVITIES FOR HOME OR SCHOOL

Food Chains

How do animals get their food?

Materials
- name tags
- colored game markers
- small plastic bags

Procedure
Play this game with ten or more people.

1. Have each player wear a tag that names him or her as a grasshopper, a snake, or a hawk. Scatter the game markers over a large area. The game markers are food.

2. Each round of the game is 30 seconds. In the first round, only grasshoppers play. They collect as many markers as they can and put them in their bags.

3. In the next round, only snakes play.

4. In the final round, only hawks play.

Draw Conclusions
Which animals had the most food after three rounds? Talk about your answer.

Energy Flow

How does energy flow through a food chain?

Materials
- index cards
- crayons
- pushpins
- yarn

Procedure

1. Divide the class into five groups: producers, plant eaters, plant and animal eaters, animal eaters, and decomposers.

2. Have each person in your group draw on an index card and label a kind of plant or animal that is from your group.

3. Form teams made up of one member from each group. Each team should make a food chain with the pictures. Use yarn to connect the parts on a bulletin board.

Draw Conclusions
How does energy flow through the food chain?

B61

Chapter 2 Review and Test Preparation

Vocabulary Review

Use the terms below to complete the sentences 1 through 9. The page numbers in ( ) tell you where to look in the chapter if you need help.

interact (B42) **energy**
producer (B43) **pyramid** (B50)
consumer (B43) **food web** (B54)
decomposer (B44) **predator** (B54)
food chain (B48) **prey** (B54)

1. A ___ feeds on the wastes of other living things.
2. A fish that is hunted and eaten by another consumer is called ___.
3. The path of food in an ecosystem from one living thing to another can be shown as a ___.
4. A ___ makes its own food.
5. A living thing that eats other living things is called a ___.
6. The living things in a community ___ with each other and with nonliving things.
7. Several linked food chains make up a ___.
8. A shark is a ___ because it hunts its food.
9. A model of how energy moves through a food web is called an ___.

Connect Concepts

Use the words listed below to complete the concept map.

consumer   horse
decomposer mushroom
grass   producer

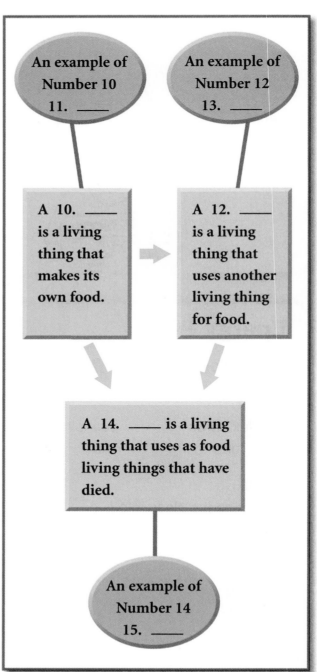

Check Understanding

Write the letter of the best choice.

16. Which is **NOT** a producer?
    - A tree
    - B flower
    - C grass
    - D bird

17. A model that shows how energy moves through a food chain is —
    - F a decomposer
    - G an ecosystem
    - H an energy pyramid
    - J a food web

18. A spider hunts and kills other animals for food. It is —
    - A prey
    - B a decomposer
    - C a predator
    - D a producer

19. Producers get their energy from —
    - F other living things
    - G the soil
    - H an energy pyramid
    - J the sun

Critical Thinking

20. A bear lives in the woods near a river. How might the bear interact with its environment to get food?

21. Where do you fit into a food chain? Draw a food chain that includes a plant or animal you ate for lunch.

Process Skills Review

22. How can you use **observation** to find out what a goat eats? How might you **infer** what the goat eats?

23. Use what you know about **models** to draw a food web that includes a bear, a water plant, berries, a big fish, a small fish, and a mouse.

Performance Assessment

Diagram a Food Web

Work with a partner. Choose an animal with which you are familiar. Draw a food web that includes the animal. Identify the producers and the consumers in each food chain. Identify predators and prey in as many food chains as you can.

UNIT B
Unit Project Wrap Up

Here are some ideas for ways to wrap up your unit project.

Display at a Science Fair
Display your mobile and cards in a school science fair. You may want to include books or other resources that tell about your ecosystem.

Draw a Mural
Show your ecosystem in a mural. Include all the living and nonliving things you saw. Label each one.

Make a Terrarium
Find samples from your ecosystem to make a terrarium. Write about how the terrarium is a model ecosystem and how the living and nonliving things in the ecosystem interact.

Investigate Further
How could you make your project better? What other questions do you have about ecosystems? Plan ways to find answers to your questions. Use the Science Handbook on pages R2–R9 for help.

References

Science Handbook
Planning an Investigation — R2
Using Science Tools — R4
   Using a Hand Lens — R4
   Using a Thermometer — R4
   Caring for and Using a Microscope — R5
   Using a Balance — R6
   Using a Spring Scale — R6
   Measuring Liquids — R7
   Using a Ruler or Meterstick — R7
   Using a Timing Device — R7
   Using a Computer — R8

Glossary — R10

Index — R18

Planning an Investigation

When scientists observe something they want to study, they use scientific inquiry to plan and conduct their study. They use science process skills as tools to help them gather, organize, analyze, and present their information. This plan will help you work like a scientist.

Step 1—Observe and ask questions.

Which food does my hamster eat the most of?

- Use your senses to make observations.
- Record a question you would like to answer.

Step 2—Make a hypothesis.

My hypothesis: My hamster will eat more sunflower seeds than any other food.

- Choose one possible answer, or hypothesis, to your question.
- Write your hypothesis in a complete sentence.
- Think about what investigation you can do to test your hypothesis.

Step 3—Plan your test.

I'll give my hamster equal amounts of three kinds of foods, then observe what she eats.

- Write down the steps you will follow to do your test. Decide how to conduct a fair test by controlling variables.
- Decide what equipment you will need.
- Decide how you will gather and record your data.

SCIENCE HANDBOOK

Step 4—Conduct your test.

I'll repeat this experiment for four days. I'll meaure how much food is left each time.

- Follow the steps you wrote.
- Observe and measure carefully.
- Record everything that happens.
- Organize your data so that you can study it carefully.

Step 5—Draw conclusions and share results.

My hypothesis was correct. She ate more sunflower seeds than the other kinds of foods.

- Analyze the data you gathered.
- Make charts, graphs, or tables to show your data.
- Write a conclusion. Describe the evidence you used to determine whether your test supported your hypothesis.
- Decide whether your hypothesis was correct.

Investigate Further

I wonder if there are other foods she will eat . . .

Using Science Tools

Using a Hand Lens

1. Hold the hand lens about 12 centimeters (5 in.) from your eye.
2. Bring the object toward you until it comes into focus.

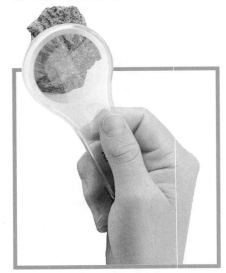

Using a Thermometer

1. Place the thermometer in the liquid. Never stir the liquid with the thermometer. Don't touch the thermometer any more than you need to. If you are measuring the temperature of the air, make sure that the thermometer is not in line with a direct light source.
2. Move so that your eyes are even with the liquid in the thermometer.
3. If you are measuring a material that is not being heated or cooled, wait about two minutes for the reading to become stable, or stay the same. Find the scale line that meets the top of the liquid in the thermometer, and read the temperature.
4. If the material you are measuring is being heated or cooled, you will not be able to wait before taking your measurements. Measure as quickly as you can.

Caring for and Using a Microscope

Caring for a Microscope

- Carry a microscope with two hands.
- Never touch any of the lenses of a microscope with your fingers.

Using a Microscope

1. Raise the eyepiece as far as you can using the coarse-adjustment knob. Place your slide on the stage.
2. Start by using the lowest power. The lowest-power lens is usually the shortest. Place the lens in the lowest position it can go to without touching the slide.
3. Look through the eyepiece, and begin adjusting it upward with the coarse-adjustment knob. When the slide is close to being in focus, use the fine-adjustment knob.
4. When you want to use a higher-power lens, first focus the slide under low power. Then, watching carefully to make sure that the lens will not hit the slide, turn the higher-power lens into place. Use only the fine-adjustment knob when looking through the higher-power lens.

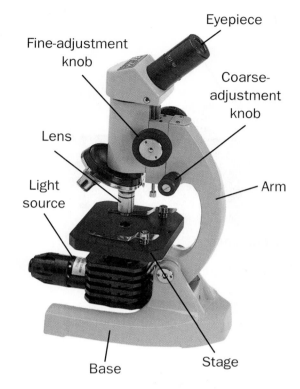

A Light Microscope

You may use a Brock microscope. This sturdy microscope has only one lens.

1. Place the object to be viewed on the stage.
2. Look through the eyepiece, and raise the tube until the object comes into focus.

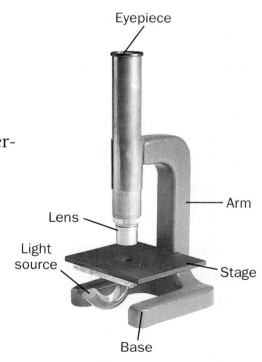

A Brock Microscope

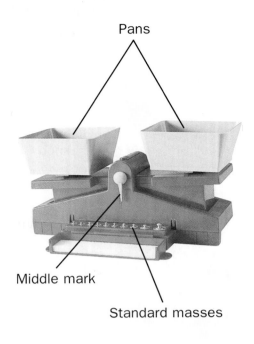

Using a Balance

1. Look at the pointer on the base to make sure the empty pans are balanced. Place the object you wish to measure in the left-hand pan.
2. Add the standard masses to the other pan. As you add masses, you should see the pointer move. When the pointer is at the middle mark, the pans are balanced.
3. Add the numbers on the masses you used. The total is the mass in grams of the object you measured.

Using a Spring Scale

Measuring an Object at Rest

1. Hook the spring scale to the object.
2. Lift the scale and object with a smooth motion. Do not jerk them upward.
3. Wait until any motion of the spring comes to a stop. Then read the number of newtons from the scale.

Measuring an Object in Motion

1. With the object resting on a table, hook the spring scale to it.
2. Pull the object smoothly across the table. Do not jerk the object.
3. As you pull, read the number of newtons you are using to pull the object.

SCIENCE HANDBOOK

Measuring Liquids

Beaker **Graduate**

1. Pour the liquid you want to measure into a measuring container. Put your measuring container on a flat surface, with the measuring scale facing you.
2. Look at the liquid through the container. Move so that your eyes are even with the surface of the liquid in the container.
3. To read the volume of the liquid, find the scale line that is even with the surface of the liquid.
4. If the surface of the liquid is not exactly even with a line, estimate the volume of the liquid. Decide which line the liquid is closer to, and use that number.

Using a Ruler or Meterstick

1. Place the zero mark or end of the ruler or meterstick next to one end of the distance or object you want to measure.
2. On the ruler or meterstick, find the place next to the other end of the distance or object.
3. Look at the scale on the ruler or meterstick. This will show the distance or the length of the object.

Using a Timing Device

1. Reset the stopwatch to zero.
2. When you are ready to begin timing, press *Start*.
3. As soon as you are ready to stop timing, press *Stop*.
4. The numbers on the dial or display show how many minutes, seconds, and parts of seconds have passed.

Using a Computer

Writing Reports

To write a report with a computer, use a word processing software program. After you are in the program, type your report. By using certain keys and the mouse, you can control how the words look, move words, delete or add words and copy them, check your spelling, and print your report.

Save your work to the desktop or hard disk of the computer, or to a floppy disk. You can go back to your saved work later if you want to revise it.

There are many reasons for revising your work. You may find new information to add or mistakes you want to correct. You may want to change the way you report your information because of who will read it.

Computers make revising easy. You delete what you don't want, add the new parts, and then save. You can also save different versions of your work.

For a science lab report, it is important to show the same kinds of information each time. With a computer, you can make a general format for a lab report, save the format, and then use it again and again.

Making Graphs and Charts

You can make a graph or chart with most word processing software programs. You can also use special software programs such as Data ToolKit or Graph Links. With Graph Links you can make pictographs and circle, bar, line, and double-line graphs.

SCIENCE HANDBOOK

First, decide what kind of graph or chart will best communicate your data. Sometimes it's easiest to do this by sketching your ideas on paper. Then you can decide what format and categories you need for your graph or chart. Choose that format for the program. Then type your information. Most software programs include a tutor that gives you step-by-step directions for making a graph or chart.

Doing Research

Computers can help you find current information from all over the world through the Internet. The Internet connects thousands of computer sites that have been set up by schools, libraries, museums, and many other organizations.

Get permission from an adult before you log on to the Internet. Find out the rules for Internet use at school or at home. Then log on and go to a search engine, which will help you find what you need. Type in keywords, words that tell the subject of your search. If you get too much information that isn't exactly about the topic, make your keywords more specific. When you find the information you need, save it or print it.

Harcourt Science tells you about many Internet sites related to what you are studying. To find out about these sites, called Web sites, look for Technology Links in the lessons in this book.

If you need to contact other people to help in your research, you can use e-mail. Log into your e-mail program, type the address of the person you want to reach, type your message, and send it. Be sure to have adult permission before sending or receiving e-mail.

Another way to use a computer for research is to access CD-ROMs. These are discs that look like music CDs. CD-ROMs can hold huge amounts of data, including words, still pictures, audio, and video. Encyclopedias, dictionaries, almanacs, and other sources of information are available on CD-ROMs. These computer discs are valuable resources for your research.

Visit the Multimedia Science Glossary to see illustrations of these words and to hear them pronounced.
www.harcourtschool.com/scienceglossary

Glossary

This Glossary contains important science words and their definitions. Each word is respelled as it would be in a dictionary. When you see the ′ mark after a syllable, pronounce that syllable with more force than the other syllables. The page number at the end of the definition tells where to find the word in your book. The boldfaced letters in the examples in the Pronunciation Key that follows show how these letters are pronounced in the respellings after each glossary word.

PRONUNCIATION KEY

a	add, map	m	move, seem	u	up, done
ā	ace, rate	n	nice, tin	û(r)	burn, term
â(r)	care, air	ng	ring, song	yo͞o	fuse, few
ä	palm, father	o	odd, hot	v	vain, eve
b	bat, rub	ō	open, so	w	win, away
ch	check, catch	ô	order, jaw	y	yet, yearn
d	dog, rod	oi	oil, boy	z	zest, muse
e	end, pet	ou	pout, now	zh	vision, pleasure
ē	equal, tree	o͝o	took, full	ə	the schwa, an unstressed vowel representing the sound spelled
f	fit, half	o͞o	pool, food		a in above
g	go, log	p	pit, stop		e in sicken
h	hope, hate	r	run, poor		i in possible
i	it, give	s	see, pass		o in melon
ī	ice, write	sh	sure, rush		u in circus
j	joy, ledge	t	talk, sit		
k	cool, take	th	thin, both		
l	look, rule	th	this, bathe		

Other symbols:
- • separates words into syllables
- ′ indicates heavier stress on a syllable
- ′ indicates light stress on a syllable

Multimedia Science Glossary: www.harcourtschool.com/scienceglossary

GLOSSARY

absorption [ab•sôrp′shən] The stopping of light **(F40)**

amphibian [am•fib′ē•ən] An animal that begins life in the water and moves onto land as an adult **(A50)**

anemometer [an′ə•mom′ə•tər] An instrument that measures wind speed **(D40)**

asteroid [as′tər•oid] A chunk of rock that orbits the sun **(D64)**

atmosphere [at′məs•fir′] The air that surrounds Earth **(D30)**

atom [at′əm] The basic building block of matter **(E16)**

axis [ak′sis] An imaginary line that goes through the North Pole and the South Pole of Earth **(D68)**

barrier island [bar′ē•ər i′lənd] A landform; a thin island along a coast **(C35)**

bird [bûrd] An animal that has feathers, two legs, and wings **(A45)**

canyon [kan′yən] A landform; a deep valley with very steep sides **(C35)**

chemical change [kem′i•kəl chānj′] A change that forms different kinds of matter **(E46)**

chlorophyll [klôr′ə•fil′] The substance that gives plants their green color; it helps a plant use energy from the sun to make food **(A20)**

clay [klā] A type of soil made up of very small grains; it holds water well **(C69)**

coastal forest [kōs′təl fôr′ist] A thick forest with tall trees that gets a lot of rain and does not get very warm or cold **(B15)**

comet [kom′it] A large ball of ice and dust that orbits the sun **(D64)**

community [kə•myoo′nə•tē] All the populations of organisms that live in an ecosystem **(B7)**

condensation [kon′dən•sā′shən] The changing of a gas into a liquid **(D17)**

conductor [kən•duk′tər] A material in which thermal energy moves easily **(F15)**

coniferous forest [kō•nif′ər•əs fôr′ist] A forest in which most of the trees are conifers (cone-bearing) and stay green all year **(B16)**

conservation [kon′ser•vā′shən] The saving of resources by using them carefully **(C76)**

R11

constellation [kon′stə•lā′shən] A group of stars that form a pattern **(D84)**

consumer [kən•sōōm′ər] A living thing that eats other living things as food **(B43)**

contour plowing [kon′tōōr plou′ing] A type of plowing for growing crops; creates rows of crops around the sides of a hill instead of up and down **(C76)**

core [kôr] The center of the Earth **(C8)**

crust [krust] The solid outside layer of the Earth **(C8)**

deciduous forest [dē•sij′ōō•əs fôr′ist] A forest in which most of the trees lose and regrow their leaves each year **(B13)**

decomposer [dē′kəm•pōz′er] A living thing that breaks down dead organisms for food **(B44)**

desert [dez′ərt] An ecosystem where there is very little rain **(B20)**

earthquake [ûrth′kwāk′] The shaking of Earth's surface caused by movement of the crust and mantle **(C48)**

ecosystem [ek′ō•sis′təm] The living and nonliving things in an environment **(B7)**

energy [en′ər•jē] The ability to cause change **(F6)**

energy pyramid [en′ər•jē pir′ə•mid] A diagram that shows that the amount of useable energy in an ecosystem is less for each higher animal in the food chain **(B50)**

environment [in•vī′rən•mənt] The things, both living and nonliving, that surround a living thing **(B6)**

erosion [i•rō′zhən] The movement of weathered rock and soil **(C42)**

estuary [es′chōō•er′•ē] A place where fresh water from a river mixes with salt water from the ocean **(D12)**

evaporation [ē•vap′ə•rā′shən] The process by which a liquid changes into a gas **(D17, E18)**

fish [fish] An animal that lives its whole life in water and breathes with gills **(A52)**

flood [flud] A large amount of water that covers normally dry land **(C50)**

food chain [fōōd′ chān′] The path of food from one living thing to another **(B48)**

food web [fōōd′ web′] A model that shows how food chains overlap **(B54)**

Multimedia Science Glossary: **www.harcourtschool.com/scienceglossary**

GLOSSARY

force [fôrs] A push or a pull **(F58)**

forest [fôr′ist] An area in which the main plants are trees **(B12)**

fossil [fos′əl] Something that has lasted from a living thing that died long ago **(C20)**

fresh water [fresh′ wôt′ər] Water that has very little salt in it **(B26)**

front [frunt] A place where two air masses of different temperatures meet **(D37)**

gas [gas] A form of matter that does not have a definite shape or a definite volume **(E12)**

germinate [jûr′mə·nāt′] When a new plant breaks out of the seed **(A13)**

gills [gilz] A body part found in fish and young amphibians that takes in oxygen from the water **(A51)**

glacier [glā′shər] A huge sheet of ice **(C44)**

gravity [grav′i·tē] The force that pulls objects toward each other **(F62)**

groundwater [ground′wôt′ər] A form of fresh water that is found under Earth's surface **(D8)**

habitat [hab′ə·tat′] The place where a population lives in an ecosystem **(B7)**

heat [hēt] The movement of thermal energy from one place to another **(F8)**

humus [hyōō′məs] The part of the soil made up of decayed parts of once-living things **(C62)**

igneous rock [ig′nē·əs rok′] A rock that was once melted rock but has cooled and hardened **(C12)**

inclined plane [in·klīnd′ plān′] A simple machine made of a flat surface set at an angle to another surface **(F71)**

inexhaustible resource [in′eg·zôs′tə·bəl rē′sôrs] A resource such as air or water that can be used over and over and can't be used up **(C94)**

inherit [in·her′it] To receive traits from parents **(A38)**

insulator [in′sə·lāt′ər] A material in which thermal energy does not move easily **(F15)**

interact [in′tər·akt′] When plants and animals affect one another or the environment to meet their needs **(B42)**

landform [land′fôrm′] A natural shape or feature of Earth's surface **(C34)**

leaf [lēf] A plant part that grows out of the stem; it takes in the air and light that a plant needs **(A7)**

lever [lev′ər] A bar that moves on or around a fixed point **(F70)**

liquid [lik′wid] A form of matter that has volume that stays the same, but can change its shape **(E12)**

loam [lōm] A type of topsoil that is rich in minerals and has lots of humus **(C70)**

lunar eclipse [loo′nər i•klips′] The hiding of the moon when it passes through the Earth's shadow **(D78)**

mammal [mam′əl] An animal that has fur or hair and is fed milk from its mother's body **(A42)**

mantle [man′təl] The middle layer of the Earth **(C8)**

mass [mas] The amount of matter in an object **(E24)**

matter [mat′ər] Anything that takes up space **(E6)**

metamorphic rock [met′ə•môr′fik rok′] A rock that has been changed by heat and pressure **(C12)**

mineral [min′ər•əl] An object that is solid, is formed in nature, and has never been alive **(C6)**

mixture [miks′chər] A substance that contains two or more different types of matter **(E41)**

motion [mō′shən] A change in position **(F59)**

mountain [moun′tən] A landform; a place on Earth's surface that is much higher than the land around it **(C35)**

nonrenewable resource [non′ri•noo′ə•bəl rē′sôrs] A resource, such as coal or oil, that will be used up someday **(C96)**

orbit [ôr′bit] The path an object takes as it moves around another object in space **(D58)**

GLOSSARY

Multimedia Science Glossary: www.harcourtschool.com/scienceglossary

phases [fāz·əz] The different shapes the moon seems to have in the sky when observed from Earth **(D76)**

photosynthesis [fōt′ō·sin′thə·sis] The food-making process of plants **(A20)**

physical change [fiz′i·kəl chānj] A change to matter in which no new kinds of matter are formed **(E40)**

physical property [fiz′i·kəl prop′ər·tē] Anything you can observe about an object by using your senses **(E6)**

plain [plān] A landform; a flat area on Earth's surface **(C35)**

planet [plan′it] A large body of rock or gas that orbits the sun **(D58)**

plateau [pla·tō′] A landform; a flat area higher than the land around it **(C35)**

population [pop′yōō·lā′shən] A group of the same kind of living thing that all live in one place at the same time **(B7)**

precipitation [prē·sip′ə·tā′shən] The water that falls to Earth as rain, snow, sleet, or hail **(D18)**

predator [pred′ə·tər] An animal that hunts another animal for food **(B54)**

prey [prā] An animal that is hunted by a predator **(B54)**

prism [priz′əm] A solid, transparent object that bends light into colors **(F44)**

producer [prə·dōōs′ər] A living thing that makes its own food **(B43)**

recycle [rē·sī′kəl] To reuse a resource to make something new **(C100)**

reflection [ri·flek′shən] The bouncing of light off an object **(F36)**

refraction [ri·frak′shən] The bending of light when it moves from one kind of matter to another **(F38)**

renewable resource [ri·nōō′ə·bəl rē′sôrs] A resource that can be replaced in a human lifetime **(C94)**

reptile [rep′tīl] A land animal that has dry skin covered by scales **(A55)**

resource [rē′sôrs] A material that is found in nature and that is used by living things **(C88)**

revolution [rev′ə·lōō′shən] The movement of one object around another object **(D68)**

rock [rok] A solid made of minerals **(C8)**

rock cycle [rok′ sī′kəl] The process in which one type of rock changes into another type of rock **(C14)**

root [rōōt] The part of a plant that holds the plant in the ground and takes in water and minerals from the soil **(A7)**

rotation [rō•tā′shən] The spinning of an object on its axis **(D68)**

S

salt water [sôlt′ wôt′ər] Water that has a lot of salt in it **(B26)**

scales [skālz] The small, thin, flat plates that help protect the bodies of fish and reptiles **(A52)**

sedimentary rock [sed′ə•men′tər•ē rok′] A rock formed from material that has settled into layers and been squeezed until it hardens into rock **(C12)**

seed [sēd] The first stage in the growth of many plants **(A12)**

seedling [sēd′ling] A young plant **(A13)**

simple machine [sim′pəl mə•shēn′] A tool that helps people do work **(F70)**

soil [soil] The loose material in which plants can grow in the upper layer of Earth **(C62)**

solar eclipse [sō′lər i•klips′] The hiding of the sun that occurs when the moon passes between the sun and Earth **(D80)**

solar system [sō′lər sis′təm] The sun and the objects that orbit around it **(D58)**

solid [sol′id] A form of matter that takes up a specific amount of space and has a definite shape **(E11)**

solution [sə•lōō′shən] A mixture in which the particles of two different kinds of matter mix together evenly **(E42)**

speed [spēd] The measure of how fast something moves over a certain distance **(F61)**

star [stär] A hot ball of glowing gases, like our sun **(D84)**

stem [stem] A plant part that connects the roots with the leaves of a plant and supports the plant above ground; it carries water from the roots to other parts of the plant **(A7)**

strip cropping [strip′ krop′ing] A type of planting that uses strips of thick grass or clover between strips of crops **(C76)**

telescope [tel′ə•skōp′] An instrument used to see faraway objects **(D88)**

temperature [tem′pər•ə•chər] The measure of how hot or cold something is **(D36)**

thermal energy [thûr′məl en′ər•jē] The energy that moves the particles in matter **(F7)**

thermometer [thûr•mom′ə•tər] A tool used to measure temperature **(F20)**

topsoil [top′soil′] The top layer of soil made up of the smallest grains and the most humus **(C63)**

Multimedia Science Glossary: www.harcourtschool.com/scienceglossary

GLOSSARY

trait [trāt] A body feature that an animal inherits; it can also be some things that an animal does **(A38)**

tropical rain forest [trop′i·kəl rān′fôr′ist] A hot, wet forest where the trees grow very tall and their leaves stay green all year **(B14)**

valley [val′ē] A landform; a lowland area between higher lands, such as mountains **(C35)**

volcano [vol·kā′nō] An opening in Earth's surface from which lava flows **(C49)**

volume [vol′yoōm] The amount of space that matter takes up **(E22)**

water cycle [wôt′ər sī′kəl] The movement of water from Earth's surface into the air and back to the surface again **(D19)**

weather [weth′ər] The happenings in the atmosphere at a certain time **(D32)**

weather map [weth′ər map′] A map that shows weather data for a large area **(D46)**

weathering [weth′ər·ing] The process by which rock is worn down and broken apart **(C40)**

weight [wāt] The measure of the pull of gravity on an object **(F62)**

wind [wind] The movement of air **(D40)**

work [wûrk] The measure of force that it takes to move an object a certain distance **(F66)**

R17

Abdominal muscles, R36
Absorption, F40
Activity pyramid, weekly plan, R22
Aerobic activities, R26–27
African elephant, A44
Agricultural chemist, A26
Agricultural extension agent, A25
Air
 animal need for, A35
 as gas, E12
 as resource, C89
Air mass, D36
Air pressure, D30
Akubo, Akira, B60
Alligators, A35, A56
Aluminum, C7
 recycling, C100, C101, C102–103
American egret, B30
Ammonites, C21
Amphibians, A50–51
Anasazi people, C10, D90
Anemometer, D40
Anemones, B27
Angler fish, B29
Animals
 adaptation of, for hunting, B44
 discovering, A58–59
 grouping, A58
 needs of, A34–38
 and people, B58–59
 sense of smell (chart), E3
 taming of, B58
 traits of, A38
 types of (chart), A31
 uses of, B58–59

Ants, C64
Anza Borrego Desert, CA, B20
Aristotle, A58–59, E30
Arteries, R40, R41
Asteroids, D64
Astronaut, D92
Astronomy, history of, D90
Atmosphere, D30
 layers of, D31
Atom(s), E16
Atomic bomb, E52
Atomic theory, E30
Axis, Earth's, D68, D71, D87

Bacon-Bercey, Joan, D50
Bacteria, B43, F9
Balance, using, R6
Balanced forces, F60
Bananas, A22
Barnacles, B27
Barn owl, A46
Barrel cactus, B21
Barrier island, C35–36
Basic light colors, F46
Bauxite, C7, C100, C103
Beaks, A46
Bears, A36
Beaufort wind scale, B40
Beaver dams, A37
Bedrock, C63
Beech tree leaf, A8
Bees, B4
Bell, Alexander Graham, F50
Biceps, R36, R37
Bicycle, safety, R12, R13
Big Dipper, D84, D86–87
Big-horned sheep, B43
Binoculars, D89
Biomedical engineer, D92

Biotite, C8
Bird(s)
 feet of, A46
 grouping, A46
 sizes of, A45
 traits of, A45
Bird banding, B34
Blackland Prairie, TX, B32
Black oak leaf, A8
Blizzards, D32
Body Systems, Human, R32–45
Brain, R44, R45
Breakage, E10
Breathing, R43
Breccia, C15
Bricks, programmable, F74–75
Buckland, William, C24
Bulbs, A14
Bullhead catfish, B30
Burning, E47, F9
Butterfly fish, A30

Cactus
 largest, A3
 parts of, B21
Camels, B59
Canopy, B15
Canyon, C36
Capillaries, R41
Carbon dioxide, A20
Cardinal, A46
Cardinal fish, A54
Cartilage, R35
Carver, George Washington, A26
Cassiopeia, D85
Casts, fossil, C21

R18

INDEX

Cat and kittens, A42–43
Cave paintings, B58
Celery, A22
Celsius scale, F21
Chalk, C6
Chameleon, A36
Cheetah cubs, A38
Chemical changes, using, E48
Chesapeake Bay, D12
Chinook, D26
Chlorophyll, A20
Circulatory system, R40–41
   caring for, R41
Civil engineer, E51
Clavicle (collar bone), R34
Clay, C69
Clouds, D17
Coal, C96
Coastal forests, B15
Cocklebur, A16
Cold front, D46
Color, E7, F44–47
   adding, F46
   of minerals, C6
Colorado River, C36
Comets, D64, D65
Communities, B7
Compact discs (CDs), F49
Composite plastic, E51
Compost, C60, F9
Compost Critters (Lavies), C65
Computer, using, R8–9
Concave lenses, F48
Condensation, D17, E41
Conductors, F15
Cone, evergreen, A12, A14
Coniferous forests, B16
Conservation, soil, C74–77
   defined, C76
Constellations, D85
Construction worker, C53

Consumers, B43, B49
Continent(s), D10–11
Continental glaciers, C44
Contour plowing, C72, C76
Cooter turtle, B30
Copernicus, D90
Copper, C7, C91
Copper sulfate, E42
Core, Earth's, C8
Corn seedling, A10
Corundum, C6
Creep, defined, C42
Crescent moon, D77
Cricket chirp thermometer, F3
Crocodiles, A56
Crop rotation, A26
Crust, Earth's, C8
Crystals, C9
Cumulus clouds, D38
Cuttings, A14

D

Dalton, John, E30
Darden, Christine, F76
Day and night, causes of, D72
Deciduous forests, B13
Decomposers, B44
Deer, A40
Delta, C43
Deltoid, R36
Democritus, E30
Desert(s)
   animals of, B22
   hot ground of, F16
   plants of, B21
   types of, B20
Devil's Tower, C32
Dew, D17

Dial thermometer, F20
Diamonds, C6, C7
Diaphragm, R42, R43
Digestive system, R38–39
   caring for, R38
Dinosauria, C24
Dinosaurs, discovering, C24–25
Dirzo, Rodolfo, A60
Distance, F61
Dogs and puppies, A43
Dolphins, B24
Doppler radar, D45
Drought-resistant plants, A24–25
Duck and ducklings, A39
Duckweed, A2
Dunes, C30, C43

E

Ear(s), caring for, R32
Ear canal, R32
Eardrum, R32
Earth, D58–59
   facts about, D61
   layers of, C8
   movement of, D68–69
   seasons of, D71
   sunlight on, D72
   surface of, C34–37, D70
   tilt on axis of, D70
   water of, D6–13
Earthquake(s), C48
   damage from, C32
   in Midwest (chart), C31
Earthquake-proof buildings, C52–53

Earthquake safety, R15
Earthworms, C64
Eastern box turtle, A55
Echidna, spiny, A44
Ecologist, C80
EcoSpun™, C107
Ecosystem(s)
    changes in, B8–9
    dangers to, B8–9
    desert, B20–23
    estuaries of, D12
    field guide, B33
    forest, B12–17
    home movies of, B22–33
    marsh, B54–55
    parts of, B7
    pond's, B7
    water, B26–31
Edison, Thomas, F50
Edison Pioneers, F50
Eggs
    bird, A45
    fish, A54
    reptile, A55
Element, E12, E30
Elodea, A18
Emerald, C6
Endangered animals (chart), B3
Endangered Species Act, B58
***Endeavour* mission,** D92
Energy
    conservation of, C101
    defined, F6
    from food, B50
    from sunlight, B49
    transfer of, B42, B48, B50
Energy pyramid, B50
Engineer, F50, F76

Environment(s)
    computer description of, B32–33
    and living things, B6
Environmental technician, D21
Equator, D11
Erosion, C42–43
Esophagus, R38, R39
Estuary(ies), D12
Evaporation, D17, D18, E18, E41
Exhaling, R43
***Extremely Weird Hunters* (Lovett),** B57
Eye(s), caring for, R32

F

Fahrenheit scale, F21
Farmer, C79
Farming
    with computers, C79
    with GPS, C78–79
Feathers, A45
Feldspar, C8
Femur, R34
Fermi, Enrico, E52
Fertilizers, C70
Fibula, R34
Film, photographic, chemical changes to, E48
Fins, A53
Fire, and thermal energy, F8
Fire safety, R14

First aid
    for bleeding, R17
    for burns, R18
    for choking, R16
    for insect bites, R19
    for nosebleeds, R18
    for skin rashes from plants, R19
First quarter moon, D77
Fish, A52–54
Fish schools, B60
Flake, plastic, C106
Flames, energy in, F6
Flexors, R36
Floods, C50, D38
Floor, forest, B15
Flowers, A14
Food
    animal need for, A36
    and bacteria, R30
    and energy, B48–51
    groups, R28
    plants use of, A20–22
    pyramid, R28
    safety tips, R31
Food chains, B48–49
Food webs, B54–57
Force(s), F58
Forest(s), types of, B12–17
Forest fires, B8
Forklift, F64
Fossils
    defined, C18, C20
    dinosaur, C24, F25
    dragonfly, C21
    fish, C3
    formation of, C20–21
    plant, C21
    types of, C20
Franklin, Benjamin, F48
Fresh water, D8–9

INDEX

Freshwater ecosystem, B26, B30
Frilled lizard, Australian, A57
Frog, metamorphosis of, A51
Fulgurite, C2
Full moon, D77
Fungi, B43

Galápagos tortoise, A56
Galilei, Galileo, D88, D90–91, F48
Gas, E11–12
  particles of, E17
Gasoline, C90
Gel, E10
Geochemist, C26
Geode, C4
Geologist, C54, D22
Germination, A13
Giant pandas, B43
Gibbous moon, D77
Gill(s), A51, A52, A53
Gilliland Elementary School, Blue Mound, TX, B22–33
Glaciers, C44–45, D8
  longest, C45
Global Learning and Observations to Benefit the Environment (GLOBE), B33
Global Positioning System (GPS), C78–79
Gneiss, C13
Gold, C6, C7
Goldfish, A53
Gorillas, A43
Graduate, E26
Granite, C9, C13

Graphite, C6, C7
Gravity, Earth's, F62
  of sun, D59
Great blue heron, A46
Great Red Spot, D62
Great Salt Lake, UT, D10
Green boa, A56
Ground cone, B2
Groundwater, D8, D18
Grouping mammals, A44
Guadeloupe bass, A52

Habitats, B7
Hail, D39
Halley's comet, D65
Hand lens, using, R4
Hardness, of minerals, C4, C6
Hawkins, Waterhouse, C25
Heart, R40, R41
  aerobic activities for, R26–27
Heat
  defined, F8
  and matter, E18
  and physical change, E38, E42
  and water forms, D16–17
High pressure, D46
Hippopotamus, B39
Hodgkin, Dorothy Crowfoot, E32
Home alone safety measures, R20–21
Hooke, Robert, A59
Hot-air balloons, D30, F2
"Hot bag," F24–25
Hubble Space Telescope, D91
Humerus, R34

Humus, C62, C68
Hurricanes, B41, D32
  record (chart), D27
Hyena, B51

Ice, D16, E11
Iceberg, D8
Icecaps, D8
Igneous rock, C12–13, C17
Iguanodon, C24
Imprints, fossil, C21
Inclined plane, F70–71
Indian paintbrush, A6
Inexhaustible resources, C94–95
Inhaling, R43
Inheritance, traits, A38
Inner ear, R32
Inner planets, D60–61
Inside the Earth (Cole), C9
Insulators, F15
Insulin, E32
Interaction, of plants and animals, B42
Internet, R9
Inventor, F26, F50
Iris, eye, R32
Iron, C7
  as conductor, F15
Irrigation, D7

Jackrabbit, B23
Jade plant leaf, A8

Jemison, Mae, D92
Jefferson Memorial, C16
Jupiter, D58–59
    facts about, D62

Keck telescope, D91
Kelp, B28
Kitchen, cleanliness, R31
Kits, beaver, A37
Koala, A44

Ladybug, A32
Lake(s), B30
Lake Baikal, Russia, C84
Lake Michigan, C43
Lake Saint Francis, AR, C32
Lakota people, D85
Landfills, C100, C101
Landforms, C34–37
    defined, C34
Landslides, Slumps, and
    Creep (Goodwin), C45
Langmuir, Charles, C26
Large intestine, R38
Lasers, F49
    and lightning, D39
Latimer, Lewis Howard, F50
Lava, C46, E18
Lava temperatures, C3
Layers, in rain forest, B14–15
Leaves, A7
    shapes of, A8
Lemon tree, A18
Lens, eye, R32
Lenses, F48
Lever, F70–71

Light
    bending, F38
    bouncing, F36–37
    and color, F44–45
    direction of, F38–39
    energy, F34
    speed of, F30, F31, F38, F44
    stopping, F40
Light and optics discovery, F48
Lightning, D48–49
Limestone, C20, C68
Lippershey, Hans, D90
Liquid(s), E11–12
    measuring, R7
    particles of, E17
Live births, A42–43
    fish, A52
    reptile, A55
Liver, R38
Living things, and food, B42–45
Lizards, A56
    described, C70
Loam, C68
Lodge, beaver, A37
Look to the North: A Wolf
    Pup Diary (George), A39
Low pressure, D46
Lunar eclipse, D78–79
Lungs, A35, A42, A45, A51, A57, R42, R43
    aerobic activities for, R26–27

Machine(s)
    compound, F72
    simple, F70–73
Magma, C8, C13

Magnet, E10
Maiman, T.H., F48
Mammals, A42–44
    types of, A44
Mangrove trees, D13
Mangrove Wilderness
    (Lavies), B51
Mantell, Gideon and Mary Ann, C24
Mantle, Earth's, C8
Mapmaking, on Internet, B33
Marble, C14, C16, C92
Mars, D58–59, F58, F62
    erosion of, D22
    facts about, D61
Mass
    adding, E27
    defined, E24
    measuring, E24–25
    of selected objects, E25
    and volume compared, E28
Matter
    appearance of, E7
    changing states of, E18
    chemical changes to, E46–49
    defined, E6
    history of classifying, E30
    measuring, E22–26
    physical changes in, E39–43
    physical properties of, E6–13
    states of, E11–12
Mayfly, B30
Measuring cup, E22
Measuring pot, E26
Measuring spoon, E26
Medeleev, Dmitri, E30
Megalosaurus, C24

INDEX

Mercury, D58–59
    facts about, D60
Mesosphere, D31
Metamorphic rock, C12–13, C16, C17
Meteor(s), D64
Meteorites, D64
Meteorologists, D32, D38, D46
Meterstick, using, R7
Microscope, A58–59, E16
    invention of, F48
    using and caring for, R5
Microwave ovens, F26
Middle ear, R32
Mineral(s)
    defined, C6
    in salt water (chart), D10
    in soil, C62, C63
    use of, C7
Mining, C90
Mirrors, F37
Mission specialist, D92
Mississippi Delta, C42
Mississippi River, C43
Mixtures, E41
    kinds of (chart), E36
Models, scientific use of, C19
Molds, fossil, C21
Montserrat Island, C49
Moon, Earth's
    and Earth interaction, D76–81
    eclipses of, D78–79
    phases of, D76–77
    rotation and revolution of, D69, D76
    weight on, F62
Moon(s), planetary, D62
Moon "rise," D74
Moose, B10
Moths, B14

Motion, F59–60
Mountain, C35
Mount Everest, Himalayas, E2
Mount Palomar observatory, D90
Mouth, R38, R39, R42, R43
Mud flows, C49
Mudslides, C50
Muscle pairs, R37
Muscovite, C8
Muscular system, R36–37
    caring for, R36
Mushroom Rock, C40

NASA, D21, D22, F76
Nasal cavity, R33
National Aeronautics and Space Administration. See NASA
National Oceanic and Atmospheric Administration, D50
National Weather Service, D45, D50
Natural gas, C97
Negative, photographic, E48
Neptune, D58–59
    facts about, D63
Nerves, R44
Nervous system, R44–45
    caring for, R45
New Madrid Fault, C32
New moon, D77
Newt(s), A50
Newton, Isaac, F48
Nice, Margaret Morse, B34
Nobel Prize in Chemistry, E32

Nobel Prize in Physics, E52
Nonliving things, in environment, B7
Nonrenewable resources, C96–97
Northern spotted owl, B15
North Star, D87
Nose, R42, R43
    caring for, R33
Nostril, R33

Obsidian, C15
Ocean, resources of, D11
Ocean ecosystems, B28–29
Ocean food web, B56
Oil
    as resource, C90
    search for, C108
Oil derrick, C90
Oil pump, C86
Oil wells, D11
Old-growth forests, C96
Olfactory bulb, R33
Olfactory tract, R33
Olympus Mons, D61
Opacity, F40
Optical fiber telephone, F48–49
Optic nerve, R32
Optics, history of, F48
Orangutan, A44
Orbit, D58
Ordering, scientific, C5
Organization for Tropical Studies, A60
Orion, D85

R23

Ornithologist, B34
Outer Banks, NC, C36
Outer ear, R32
Outer planets, D60, D62–63
Owen, Richard, C24–25
Oxygen, A35, A52
    and plants, A21
    as resource, C88, C89

Pacific Ocean, D10
Pan balance, E24
Panda, A36, B53
Paper, changes to, E39
Partial lunar eclipse, D79
Partial solar eclipse, D80
Particles
    bumping, F14
    connected, E17
    in fire, F7
    and heat, F7
    in liquids and gases, F16
    of solids, F7
Patents, F50
Peace Corps, D92
Peanuts, A26
Pecans, A26
Pelvis, R34
People and animals, history
    of, B58
Periodic table, E31
Perseus, D85
Petrologist, C26
Photosynthesis, A20–21
Physical changes,
    kinds of, E39
Physical property,
    defined, E6
Physician, D92
Physicist, E52

Pill bugs, C64
Pine cone, largest, A3
Pizza chef, F25
Pizza holder,
    as insulator, F15
Plain, C35
Planets, D58
    distance from sun, D58–59
    facts about, D57
Plankton, B60
Plant(s)
    drought-resistant, A24–25
    foodmaking, A20–21
    heights of (chart), A2
    need for light, F34
    needs of, A6–9
    parts of, A7
    seed forming, A14
    use of food by, A22–23
    and weathering, C41
Plant material, in different
    environments (chart), B38
Plastic, C90, C100
    bridges from, E50–51
    recycling of, C106–107
Plateau, C35
Pluto, D58–59
    facts about, D63
Polar bear, A42
Polaris, D87
Pollution, D8, D40, C101
    preventing, C104
Polo, Marco, F48
Ponds, B30
Populations, B7
Postage scale, E26
Potatoes, A22
Potting soil, C70
Precipitation, D8, D17,
    D18, D32
    measuring, D38
Predator, B54

Prey, B54
Prisms, F44
Producers, B43, B49
Property, defined, C4
Puffin, B46
Pull, F58
Pulley, F70–71
Pumice, C15
Pupil, eye, R32
Purple gallinule, A46
Push, F58, F60

Quadriceps, R36
Quartz, C6, C7, C16
Quartzite, C14
Quinones, Marisa, C108

Raccoon, B6
Radiation, F16
Radio telescope,
    Arecebo, D91
Radishes, A4
Radius, R34
Rain, D32
Rainbow(s), F44
    formation of, F45
Rainbow trout, A53
Rainfall, by type
    of forest, B12
Rain forests, C75
    See also Tropical rain
    forests
Rain gauges, D38
Ray, John, A58–59

INDEX

Recycling, C100–103
    plastics, C106–107
Recycling plant worker, C107
Red sandstone, C15
Reflection, F36
Refraction, F38, F44
Renewable resources, C94
Reptiles, A55–57
Resources
    conserving, C104
    daily use of Earth's, C85
    defined, C74
    kinds of, C94–97
    location of, C88–89
    under Earth's surface, C90
Respiratory system, R42–43
    caring for, R43
Retina, R32
Revolution, Earth's, D68
Rhinoceros, B3
Rib, R35
Rib cage, R34, R35
Rivers, B30, C35, C50
Rock(s)
    described, C8
    formation of, C12–15
    types of, C12–15
    use of, C16–17
    weathering of, C40–41
Rock, The (Parnall), C17
Rock cycle, C14–15
Rocky Mountains, C32
Roots, A7, A14
Rose, fragrance of, E9
Rossbacher, Lisa, D22
Rotation, Earth's, D68
Rowland, Scott, C54
Rubbing, and thermal energy, F8
Ruler, using, R7
Rusting, E47

S

Safety
    and health, R12–21
    in Science, xvi
Saguaro cactus, B22
Salamander, A50
Salt water, D10–11
Saltwater ecosystems, B26, B27–28
San Andreas Fault, CA, C48
Sand, C69
Sandstone, C15
Sapphire, C6
Satellite, D45
Saturn, D58–59
    facts about, D62
Scales, fish, A52, A53
Scavenger, B51
Schist, C13
Scissors, operation of, F72
Scorpion, B18
Screw, F70–71
Sea horse, A38
Sea otter, B52
Seasons, D68–73
    causes of, D70
Sediment, C12
Sedimentary rock, C12–13
Seed(s)
    berry, A16
    described, A12
    kinds of, A14
    mangrove, A16
    milkweed, A17
    needs of, A13
    parts of, A15
    sizes of, A15
    spreading, A16–17
    sunflower, A12
Seed coat, A15

Seedling, A13, A15
Segre, Emilio, E30
Sense(s), describing matter through, E7–10
Sense organ(s), R32–33
Serving sizes, R29
Shadows, D69, F32, F35
Shale, C20
Shape, of minerals, C6
Sharks, A52
    sense of smell of, E37
Shelter, animal need for, A37
Sidewinder, B22
Silicon chips, C16
Size, E7
Skeletal system, R34–35
    caring for, R35
Skin, R33
    layers of, R33
Skull, R34, R35
Skunk, odor of, E9
Sky watchers, D90–91
Slate, C13, C14
Small intestine, R38
Snakes, A38, A56, B22
Snow boards, D38
Snowflake porphyry, C15
Soil
    conservation of, C74–77
    formation of, C62–63
    importance of, C64–65
    life in, C64
    living things in (chart), C59
    of Mars, C59
    minerals in, A7
    parts of, C70
    pigs use of, C58
    as resource, C89
    types of, C68–71

Soil roundworms, C80
Solar eclipse, D80
Solar energy, F35
Solar system, D58–65
   structure of, D58
Solid(s), E11
   particles of, E17
Solution, E42
   to solid, E46
Sonic boom, F76
Speed, F54, F60–61
   of different objects (chart), F55
Spencer, Percy, F26
Spiders, B8
Spinal cord, R44, R45
Spine, R34
Spring scale, E26, F56, F58
   using, R6
Squash, growth of, A13
Standard masses, R6
Star(s)
   defined, D84
   and Earth's movement, D86–87
   observing, D84–89
   patterns, D84–85
   sizes of (chart), D55
Stars: Light in the Night, The **(Bendick),** D89
Stem, A7
Sternum, R35
Stickleback, A54
Stomach, R38, R39
Stonehenge, D90
Stored food, seed, A15
Storm clouds, D28
Storm safety, R15
Stratosphere, D31
Stratus clouds, D38
Strawberries, A22
Streams, B30
Stretches, R24–25
Strip cropping, C26
Strip mining, C90, C103
Subsoil, C63
Sugar, plant, A21
Sun
   eclipse of, D80
   heat from, F16
   light energy of, A20–21
   as star, D59
Sun and Moon, The **(Moore),** D81
Sweet gum leaf, A8
Sweet potatoes, A26

T

Tadpoles, A49, A51
Taklimakan desert, China, B20
Taste buds, R33
Tayamni, D85
Teacher, B33
Technetium, E30
Telescopes, D56, D82, D88, D91, D95
Temperature, D32, E8
   measuring, D36, F20–23
   by type of forest, B12
   various measurements of, F20–21
Tendon(s), R37
Texas bluebonnets, A6
Thermal energy, D16, F6–11, F24
   controlling, F22
   movement of, F14–17
   producing, F9
   use of, F10
Thermal retention, F24–25
Thermal vents, A59
Thermometer(s), F18, F21
   using, R4
   working of, F20
Thermosphere, D31
Thermostat, F23
   working of, F22
Third quarter moon, D77
Thunderhead, D28
   composition of, D39
Tibia, R34
Tilling, C76
Timing device, using, R7
Toads, A50
Tongue, R33
Tools, of measurement, E26
Topsoil, C63, C70
Tornadoes, B41, D33, D39
Tortoises, A56
Total lunar eclipse, D79
Total solar eclipse, D80
Toy designer, F75
Trachea, R42
Traits, animal, A38, A44
Translucency, F40
Transparency, F40
Trash, C101
Tree frog, A48
Triceps, R36, R37
Triton, temperatures on, D55
Tropical ecologist, A60
Tropical fish, B29
Tropical rain forests, B14–15
   See also Rain forests
Troposphere, D31
Tubers, A14
Turtles, A56, B48
Tuskegee Institute, AL, A26
Tyrannosaurus rex, C18

INDEX

Ulna, R34
Umpqua Research Company, D21
Understory, B15
Uranus, D58–59
   facts about, D63
Ursa Major, D84

Valles Marineris, Mars, D54
Valley, C35
Valley glaciers, C44
Valve, R41
Veins, R40, R41
Venus, D58–59
   facts about, D60
Verdi **(Cannon),** A57
Very Large Array telescope, D95
Volcanoes, C13, C46, C49–50, E18
   Hawaiian, C54
   types of, C54
Volume
   defined, E22
   and mass compared, E28
   measuring, E22–23

Wall, Diana, C80
Warm front, D46
Waste (chart), C100
Water
   amounts of salt and fresh, D7
   forms of, D16–17
   importance of, D6–7
   as resource, C88
   in space, D21
   uses of, D3
   wasting, D2
   and weathering, C41
Water cycle, D16–19
Water filters, D20–21
Water hole, African, A35
Water treatment plant, D11
Water vapor, D16, D17
Weather, D32
   gathering data, D44–45
   measuring, D36–41
Weather balloons, D45
Weather forecasting, D44–47
Weather fronts, D37, D46
Weather maps, D46
Weather researcher, D49
Weather satellites, D44
Weather station, D44
   symbols used by, D46, D47
Weather vanes, D34
Weathering, C40–41
Weaverbird, A45
Wedge, F70, F72
Weighing scale, E20
Weight, F62
Wells, C90
West Chop Lighthouse, MA, C31
Wetness, E8
Whales, A44
Wheel and axle, F70, F72
White light, F46
   colors of, F45
Wildfires **(Armbruster),** B9
Wind, D32
   measuring, D40
Wings, A45
Women in Science and Engineering (WISE) Award, F76
Work, F66–67
Workout, guidelines for, R23–25
World Health Organization, D20–21
Wright, James, E37

Xeriscaping, A25

Yellowstone National Park, B8
Yellow tang, A53

Zinc, C7

Photography Credits - Page placement key: (t) top, (c) center, (b) bottom, (l) left, (r) right, (bg) background, (i) inset

Cover Background, Charles Krebs/Tony Stone Images; Inset, Jody Dole.

Table of Contents - iv (bg) Thomas Brase/Tony Stone Images; (i) Denis Valentine/The Stock Market; v (bg) Derek Redfeam/The Image Bank; (i) George E. Stewart/Dembinsky Photo Association; vi (bg) Richard Price/FPG International; (i) Martin Land/Science Photo Library/Photo Researchers; vii (bg) Pal Hermansen/Tony Stone Images; (i) Earth Imaging/Tony Stone Images; viii (bg) Steve Barnett/Liaison International; (i) StockFood America/Lieberman; ix (bg) Simon Fraser/Science Photo Library/Photo Researchers; (i) Nance Trueworthy/Liaison International.

Unit A - A1 (bg) Thomas Brase/Tony Stone Images; (i) Denis Valentine/The Stock Market; A2-A3 (bg) Joe McDonald/Bruce Coleman; A3 (i) Marilyn Kazmers/Deminsky Photo Associates; A4 Ed Young/AGStock USA; A6 (l) Anthony Edgeworth/The Stock Market; (r) Chris Vincent/The Stock Market; A6-A7 (bg) Barbara Gerlach/Dembinsky Photo Associates; A7 (c) Wendy W. Cortesi; A8 (t) Runk/Schoenberger/Grant Heilman Photography; (c) Runk/Schoenberger/Grant Heilman Photography; (bl) Renee Lynn/Photo Researchers; (br) Dr. E.R. Degginger/Color-Pic; A9 Runk/Schoenberger/Grant Heilman Photography; A10 Runk/Schoenberger/Grant Heilman Photography; A12 (l) Bonnie Sue/Grant Heilman/Photo Researchers; (li) Klaus Paysan/Peter Arnold, Inc.; (r) Runk/Schoenberger/Grant Heilman Photography; (ri) Runk/Schoenberger/Grant Heilman Photography; A13 (tr) Ed Young/AgStock USA; (b) Dr. E. R. Degginger/Color-Pic; A14 (tr) Richard Shiell/Dembinsky Photo Associates; (bl) Robert Carr/Bruce Coleman, Inc.; (br) Scott Sinklier/AGStock USA; A16 (t) Thomas D. Mangelsen/Peter Arnold, Inc.; (c) E.R. Degginger/Natural Selection Stock Photography; (bl) Randall B. Henne/Dembinsky Photo Associates; (br) Stan Osolinski/Dembinsky Photo Associates; (I) Scott Camazine/Photo Researchers; A17 William Harlow/Photo Researchers; A18 Christi Carter/Grant Heilman Photography; A20 Runk/Schoenberger/Grant Heilman Photography; A22 (t) DiMaggio/Kalish/The Stock Market; (cl) Jan-Peter Lahall/Peter Arnold, Inc.; (br) Holt Studios/Nigel Cattlin/Photo Researchers; A23 Robert Carr/Bruce Coleman; A24 Richard Shiell; A25 J. Sapinsky/The Stock Market; A26 (tr) Corbis; A30-A31 (bkgd) Art Wolfe/Tony Stone Images; A31 (cr) Astrid & Hanns Frieder Michler/Science Photo Library/Photo Researchers; A32 (bl) Rosemary Calvert/Tony Stone Images; (1) Ralph A. Reinhold/Animals Animals; (2) Johnny Johnson/ Tony Stone Images; (3) Mike Severns/ Tom Stack & Associates; (4) Fred Whitehead/Animals Animals; (5) Art Wolfe/ Tony Stone Images; (6) J.C. Stevenson/Animals Animals; A34-A35 Doug Perrine/Innerspace Visions; A35 (t) Ronald Hellstrom/Bruce Coleman, Inc.; (br) Stan Osolinski/Tony Stone Images; A36 (tr) Mike Severns/Tony Stone Images; (lc) Kevin Schafer Photography; (bl) Marilyn Kazmers/Peter Arnold, Inc.; (br) Keren Su/Tony Stone Images; A38 (t) Rudie Kuiter/Innerspace Visions; (c) Fred Bruemmer/Peter Arnold, Inc.; (b) Art Wolfe/Tony Stone Images; A39 Phil A. Dotson/Photo Researchers; A40 Brian Stablyk/Tony Stone Images; A43 (t) Paul Metzger/Photo Researchers; (b) Frans Lanting/Minden Pictures; A44 (t) Stephen Dalton/Photo Researchers; (c) Tom McHugh/Photo Researchers; (cr) Evelyn Gallardo/Peter Arnold, Inc.; (cl) The Photo Library-Sydney/Gary Lewis/Photo Researchers; (br) Francois Gohier/Photo Researchers; A45 (t) Theo Allofs/Tony Stone Images; (blue jay) Wayne Lankinen/Bruce Coleman, Inc.; (macaw) M. Mastrorillo/The Stock Market; (emperor penquin) Kjell B. Sandved/Photo Researchers; (ostrich) Leonard Lee Rue III/Photo Researchers; (bee humming bird) Robert A. Tyrrell Photography; (peacock) Tom McHugh/Photo Researchers; A46 (t) Manfred Danegger/Tony Stone Images; (cl) John Cancalosi/Peter Arnold, Inc.; (b) Bill Ivy/Tony Stone Images; (br) Stan Osolinski/The Stock Market; A48 (c) O.S.F./ Animals Animals; (b) Tim Davis/Tony Stone Images; A50 (tl) Nuridsany et Perennou/Photo Researchers; (r) E.R. Degginger/Photo Researchers; (bl) Joseph T. Collins/Photo Researchers; A52 (t) David M. Schleser/Nature's Images; (c) Andrea & Antonella Ferrari/Innerspace Visions; A52-A53 Kelvin Aitken/Peter Arnold, Inc.; A53 (t) Zig Leszczynski/Animals Animals; (c) Kelvin Aitken/Peter Arnold, Inc.; (br) Tom McHugh/Steinhart Aquarium/Photo Researchers; A54(t) Kim Taylor/Bruce Coleman, Inc.; (c) Fred Bavendam/Minden Pictures; (b) Fred Bavendam/Minden Pictures; A55 (t) Zig Leszczynski/Animals Animals; (b) Suzanne L. Collins & Joseph T. Collins/Photo Researchers; (bli) Dwight R. Kuhn; A56 (t) Jany Sauvanet/Photo Researchers; (c) G.E. Schmida/Fritz/Bruce Coleman, Inc.; A56-A57 (b) Tom & Pat Leeson/Photo Researchers; A57 Schafer & Hill/Tony Stone Images; A58 (t) Tom Brakefield/Bruce Coleman, Inc.; (c) Dr. E. R. Degginger/Color-Pic; (b) Michael Holford; A59 Emory Kristof/National Geographic Image Collection; A60 (tr) Bertha G. Gomez; (bl) Michael Fogden/bruce Coleman, Inc.

Unit B - B1 (bg) Derek Redfeam/The Image Bank; (i) George E. Stewart/Dembinsky Photo Association; B2-B3 (bg) Sven Linoblad/Photo Researchers; B2 (i) Wayne P. Armstrong; B4 Hans Pfletschinger/Peter Arnold, Inc.; B6 (I) Dwight R. Kuhn; (r) Michael Durham/ENP Images; B7 Frank Krahmer/Bruce Coleman, Inc.; B8 (tl) Jeff and Alexa Henry/Peter Arnold, Inc.; (t) Jeff and Alexa Henry/Peter Arnold, Inc.; (b) Christoph Burki/Tony Stone Images; B10 Kennan Ward/The Stock Market; B13 (all) James P. Jackson/Photo Researchers; B14 Zefa Germany/The Stock Market; B15 Janis Burger/Bruce Coleman Inc.; B16 (r) Michael Quinton/Minden Pictures; B16-B17 (b)Grant Heilman Photography; B18 J.C. Carton/Bruce Coleman, Inc.; B20 (l) Wolfgang Kaehler Photography; (r) James Randklev/Tony Stone Images; B21 Dr. E.R. Degginger/Color-Pic; B22 (l) Paul Chesley/Tony Stone Images; (r) Jeff Foott/Bruce Coleman, Inc.; (i) Jen & Des Bartlett/Bruce Coleman, Inc.; B23 Lee Rentz/Bruce Coleman, Inc.; B24 Leo De Wys Inc.; B27 (li) R.N. Mariscal/Bruce Coleman, Inc.; (b) Dr. E.R. Degginger/Color-Pic; (ri) Naitar E. Harvey, APSA/National Audubon Society/Photo Researchers; B28 Flip Nicklin/Minden Pictures; B29 (t) Norbert Wu/Peter Arnold, Inc.; (b) Norbert Wu/Peter Arnold, Inc.; B30 (t) Gary Meszaros/Bruce Coleman, Inc.; (bli) Stevan Stefanovic/Okapia/Photo Researchers; (bci) Dwight R. Kuhn; (bri) Phil Degginger/Color-Pic; B30-B31 (b) Jeff Greenberg/Photo Researchers; B32 (b) Courtesy of Jane Weaver/Parie Project/L. A. Gilillard Elementary; (ti) Globe-NASA/ Goddard Scientific Visualization Studio; B33 Derke/O'Hara/Tony Stone Images; B34 (tr) The Marjorie N. Boyer Trust; (bl) Anthony Merciece/ Photo Researchers; B38-B39 Luiz C. Marigo/Peter Arnold, Inc.; B39 (br) Roland Seitre/Peter Arnold, Inc.; B40 (l) Roy Morsch/The Stock Market; (c) Norbert Wu/Tony Stone Images; (r) Rosemary Calvert/Tony Stone Images;
B41 (bl) Stan Osolinski/The Stock Market; (c) R. Kopfle/KOPFL/Bruce Coleman; (br) Michael Durham/ENP Images; B42 (bg) Hans Reinhard/Bruce Coleman, Inc.; (li) Dwight R. Kuhn; (ri) Dr. Paul A. Zahl/Photo Researchers; B43 (t) Wolfgang Kaehler Photography; (b) Rob Hadlow/Bruce Coleman, Inc.; B44 (t) Stephen Dalton/Photo Researchers; (c) Andrew Syred/Science Photo Library/Photo Researchers; (b) Stephen Krasemann/Tony Stone Images; B46 Laurie Campbell/Tony Stone Images; B48 Dwight R. Kuhn; B49 (t) Paul E. Taylor/Photo Researchers; (c) Holt Studios/Photo Researchers; (b) Breck P. Kent/Animals Animals; B51 Mitsuaki Iwago/Minden Pictures; B52 Erwin and Peggy Bauer/Bruce Coleman, Inc.; B54-B55 Michael Durham/ENP Images; B57 Jane Burton/Bruce Coleman; B58 (cl) LASCAUX Caves II, France/Explorer, Paris/Superstock; (c) Fred Bruemmer/Peter Arnold; (bc) Tom Brakefield/Bruce Coleman; B60 (tl) Leah Edelstein-Keshet/University of British Columbia; (bl) Fred McConnaughey/Photo Researchers.

Unit C Other - C1(bg) Richard Price/FPG International; (i) Martin Land/Science Photo Library/Photo Researchers; C2-C3 (bg) E. R. Degginger; C2 (bc) A. J. Copley/Visuals Unlimited; C3 (ri) Paul Chesley/Tony Stone Images; C4 (b) The Natural History Museum, London; C6 (tl), (ct), (cb) Dr. E. R. Degginger/Color-Pic; (r) E. R. Degginger/Bruce Coleman, Inc.; (bl) Mark A. Schneider/Dembinsky Photo Associates; C6-C7 (b) Chromosohm/Joe Sohm/Photo Researchers; C7 (tr) Blair Seitz/Photo Researchers; (tri), (bli) Dr. E. R. Degginger/Color-Pic; C8 (t) Barry Runk/Grant Heilman Photography; (ct) Dr. E. R. Degginger/Color-Pic; (b) Dr. E. R. Degginger/Color-Pic; (cb) Barry L. Runk/Grant Heilman Photography; C8-9 (t) Robert Pettit/Dembinsky Photo Associates; C10 Tom Bean/Tom & Susan Bean, Inc.; C12 Jim Steinberg/Photo Researchers; C12-C13 (bg) G. Brad Lewis/Photo Resource Hawaii; C14 (l), (tr) Dr. E. R. Degginger/Color-Pic; C14 (br) Aaron Haupt/Photo Researchers; C15 (tl) Robert Pettit/Dembinsky Photo Associates; (tr) Charles R. Belinky/Photo Researchers; (bl), (bc), (br) Dr. E. R. Degginger/Color-Pic; C16 (t) Roger Du Buisson/The Stock Market; (c) Jay Mallin Photos; C16-C17 (b) Ed Wheeler/The Stock Market; C18 Stephen Wilkes/The Image Bank; C21 (t) William E. Ferguson; (bl) Kerry T. Givens/Bruce Coleman, Inc.; (br) Joy Spurr/Bruce Coleman, Inc.; C22 (t) AP Photo/Dennis Cook; (b) M. Timothy O'Keefe/Bruce Coleman, Inc.; C24 (t) Francois Gohier/Photo Researchers; (b) The National History Museum/London; C25 Stan Osolinski; C26 (tr) Jean Miele/Lamont-Doherty Earth Observatory of Columbia University; C30-C31 (bg) John Warden/Tony Stone Images; C31 (bl) Harold Naideau/The Stock Market; C32 G. Alan Nelson/Dembinsky Photo Associates; C33 (b) Superstock; C34-35 (b) Darrell Gulin/Dembinsky Photo Associates; C35 (t) Michael Hubrich/Dembinsky Photo Associates; (c) Mark E. Gibson; C36 (t) Breck P. Kent/Earth Scenes; C36-37 (b) Paraskevas Photography; C38 Mark E. Gibson; C40 (bl) Dr. E.R. Degginger/Color-Pic; C40 (bc) Mark A. Schneider/Dembinsky Photo Associates; C40-C41 (br-b) Rod Planck/Dembinsky Photo Associates; C41 (t) Michael Hubrich/Dembinsky Photo Associates; (c) John Gerlach/Dembinsky Photo Associates; C42 (t) Georg Gerster/Photo Researchers; (c) NASA Photo/Grant Heilman Photography; C42-C43 (b) C.C. Lockwood/Earth Scenes; C43 (t) Mark E. Gibson; C46 Ken Sakamoto/Black Star; C48 (l) David Parker/SPL/Photo Researchers; (l) AP/Wide World Photos; C49 (l) AP/Wide World Photos; (r) AP Photo/Wide World Photos; (bl) Will & Deni McIntyre/Photo Researchers; C50-C51 AP/Wide World Photos; C52 (l) George Hall/Woodfin Camp & Associates; (i) Laura Riley/Bruce Coleman; C53 J. Aronovsky/Zuma Images/The Stock Market; C54 (tr) Courtesy of Scott Rowland; (bl) Dennis Oda/Tony Stone Images; C58-C59 (bg) Lynn M. Stone/Bruce Coleman, Inc.; C59 (br) NASA; C60 Ann Duncan/Tom Stack & Associates; C63 (all) Bruce Coleman, Inc.; C66 Grant Heilman/Grant Heilman Photography; C68-C69 (b) Gary Irving/Panoramic Images; C68 (I) Barry L. Runk/Grant Heilman Photography; C69 (li), (ri) Barry L. Runk/Grant Heilman Photography; C70-C71 (b) Larry Lefever/Grant Heilman Photography; C72 Andy Sacks/Tony Stone Images; C74 USDA - Soil Conservation Service; C74-C75 (b) Dr. E.R. Degginger/Color-Pic; C75 (t) James D. Nations/D. Donne Bryant; (tli) Gunter Ziseler/Peter Arnold, Inc.; (tri) S.AM./Wolfgang Kaehler Photography; (bli) Walter H. Hodge/Peter Arnold, Inc.; (bri) Jim Steinberg/Photo Researchers; C76 (t) Thomas Hovland from Grant Heilman Photography; (b) B.W. Hoffmann/AGStock USA; C78 (b) Randall B. Henne/Dembinsky Photo Associates; (I) Russ Munn/AgStock USA; C79 Bruce Hands/Tony Stone Images; C80 (tr) Courtesy of Diana Wall, Colorado State University; (bl) Oliver Mickes/Ottawa/Photo Researchers; C84-C85 (bg) Kirby, Richar OSF/Earth Scenes; C86 Bob Daemmrich/Bob Daemmrich Photography, Inc.; C88 (l) Peter Correz/Tony Stone Images; (c) Mark E. Gibson; C88-C89(b) Bill Lea/Dembinsky Photo Associates; C90 (t) Chris Rogers/Rainbow/PNI; (c) Yoav Levy/Phototake/PNI; C91 Rob Badger Photography; C92 (bl) Christie's Images, London/Superstock; (br) Jeff Greenberg / Photo Researchers; (bc) Joyce Photographics / Photo Researchers; (tr) Mary Ann Kulla/ The Stock Market; C93 (bl) Alan L. Detrick / Photo Researchers; (bc) Archive Photos; (br) David Barnes /The Stock Market; (tr) Gary Retherford/ Photo Researchers; C94-C95 Jeff Greenberg/Visuals Unlimited; C95 (tl) Wolfgang Fischer/Peter Arnold, Inc.; (tr) Wolfgang Fischer/Peter Arnold, Inc.; (c) Craig Hammell/The Stock Market; C96 (t) Barbara Gerlach/Dembinsky Photo Associates; (b) Brownie Harris/The Stock Market; C97 Chris Rogers/The Stock Market; C98 Michael A. Keller/The Stock Market; C100-C101 (b) Ray Pfortner/Peter Arnold, Inc.; C103 William E. Ferguson; C104-C105 Blaine Harrington III/The Stock Market; C106 James King-Holmes/Science Photo Library/ Photo Researchers; C107 (bl) Wellman Fibers Industry; (r) Gabe Palmer/The Stock Market; C108 (tr) Susan Sterner/HRW; (bl) Kristin Finnegan/Tony Stone Images.

Unit D Other - D1(bg) Pal Hermansen/Tony Stone Images; (i) Earth Imaging/Tony Stone Images; D2-D3 (bg) Zefa Germany/The Stock Market; D3 (tr) Michael A. Keller/The Stock Market; (br) Steven Needham/Envision; D4 J. Shaw/Bruce Coleman, Inc.; D5 (b) NASA; D6 (l) Yu Somov; S. Korytnikov/Sovfoto/Eastfoto/PNI; (r) Christopher Arend/Alaska Stock Images/PNI; D7 Grant Heilman Photography; D8-D9 Dr. Eckart Pott/Bruce Coleman, Inc.; D10 N.R. Rowan/The Image Works; D12-D13 Mike Price/Bruce Coleman, Inc.; D13 David Job/Tony Stone Images; D14 John Beatty/Tony Stone Images; D17 (l) Grant Heilman Photography; (r) Darrell Gulin/Tony Stone Images; D20 NASA; D21 Ben Osborne/Tony Stone Images; D22 (tr) Polytehnic State University; (b) NASA; D26-D27 Andrea Booher/Tony Stone Images; D27 NASA/Science Photo Library/Photo Researchers; D28 Joe Towers/The Stock Market; D30 Rich Iwasaki/Tony Stone Images; D32 (t) Peter Arnold; (c) Warren Faidley/International Stock Photography; (b) Stephen Simpson/FPG International; D33 E.R. Degginger/Bruce Coleman, Inc.; D34 Ray Pfortner/Peter Arnold, Inc.; D37 (t) Ralph H. Wetmore, II/Tony Stone Images; (b) Joe McDonald/Earth Scenes; D38 (c) Tom Bean; (b) Adam Jones/Photo Researchers; D42 Warren Faidley/International Stock Photography; D44 (l) Superstock; (r) David Ducros/Science Photo Library/Photo Researchers; D45 (both) © 1998 AccuWeather; D45 New Scientist Magazine; D49 Dwayne Newton/PhotoEdit; D50 (tr) Courtesy June Bacon-Bercey; (bl) David R. Frazier/Photo Researchers; D54-D55 David Hardy/Science Photo Library/Photo Researchers; D55 (tc) European Space Agency/Science Photo Library/Photo Researchers; D60 (t) U.S. Geological Survey/Science Photo Library/Photo Researchers; D60 (b) NASA; D60, D61, D62 (bg) Jerry Schad/Photo Researchers; D61 (t) National Oceanic and Atmospheric Administration; D61 (b) David Crisp and the WFPC2 Science Team (Jet Propulsion Laboratory/California Institute of Technology); D62 (t), (b) NASA; D63 (t) Erich Karkoschka (University of Arizona Lunar & Planetary Lab) and NASA; (c) NASA; (b) Nasa/Science Source/Photo Researchers; D64 J. Spurr/Bruce Coleman, Inc.; D65 Royer, Ronald/Science Photo Library/Photo Researchers; D66 Renee Lynn/Photo Researchers; D69 (l) Dr. E. R. Degginger/Color-Pic; (r) Dr. E. R. Degginger/Color-Pic; D72 (t) Joseph Nettis/Photo Researchers; (b) John Elk III/Bruce Coleman, Inc.; D74 NASA; D77 (all) Telegraph Colour Library/FPG International; D78-D79 Margaret Miller/Photo Researchers; D79 (t) Pekka Parviainen/Science Photo Library/Photo Researchers; (b) George East/Science Photo Library/Photo Researchers; D80 Dr. Fred Espenak/Science Photo Library/Photo Researchers; D82 The Granger Collection, New York; D88 Merritt Vincent/PhotoEdit; D90 (l) Rob Talbot/ Tony Stone Images; (r) Stephen Graham/Dembinsky Photo Associates; D91 NASA; D92 (both) NASA.

Unit E Other - E1(bg) Steve Barnett/Liaison International; (i) StockFood America/Lieberman; E2-E3 Chris Noble/Tony Stone Images; E3 Kent & Donna Dannen; E6 John Michael/International Stock Photography; E7 (r) R. Van Nostrand/Photo Researchers; E8 (tr) Mike Timo/Tony Stone Images; E9 (t) Daniel J. Cox/Tony Stone Images; (ti) Goknar/Vogue/Superstock; E10 (br) John Michael/International Stock Photography; (bc) William Cornett/Image Excellence Photography; E11 (tr) Lee Foster/FPG International; (br) Paul Silverman/Fundamental Photographs; E12 (t) William Johnson/Stock, Boston; E12-E13 (b) Robert Finken/Photo Researchers; E14 S.J. Krasemann/Peter Arnold, Inc.; E16 (ri) Dr. E. R. Degginger/Color-Pic; E17 (l) Charles D. Winters/Photo Researchers; (br) Spencer Gran/PhotoEdit; E18-E19 Peter French/Pacific Stock; E20 J. Sebo/Zoo Atlanta; E25 (cr) Robert Pearcy/Animals Animals; E25 (cl) Ron Kimball Photography; E26 (tr) Jim Harrison/Stock, Boston; E32 (tr) Corbis; (bl) Alfred Pasieka/Science Photo Library/Photo Researchers; E36-E37 (bg)Dr. Dennis Kunkel/Phototake; E38 Robert Ginn/PhotoEdit; E42 (tl), (tr) Dr. E. R. Degginger/Color-Pic; (bl) Tom Pantages; (br) Tom Pantages; E44 Chip Clark; E46 (l) Tom Pantages; (r), (c) Tom Pantages; E47 (bl) John Lund/Tony Stone Images; E50 John Gaudio; E51 Michael Newman/Photo Edit; E52 (tr) Los Alamos National Laboratory/Photo Researchers; (bl) US Army White Sands Missile Range.

Unit F - F1 (bg) Simon Fraser/Science Photo Library/Photo Researchers; (i) Nance Trueworthy/Liaison International; F2 (bg) April Riehm; (tr) M. W. Black/Bruce Coleman, Inc.; F6 (bl) Mary Kate Denny/PhotoEdit; (br) Gary A. Conner/PhotoEdit; F7 (bl) Camerique, Inc./The Picture Cube; (br) Stephen Saks/The Picture Cube; F8 (cr) Pat Field/Bruce Coleman, Inc.; (bl) Mark E. Gibson; F9 (l) Dr. E.R. Degginger/Color-Pic; (r)Ryan and Beyer/Allstock/PNI; F10 (b) John Running/Stock Boston; (cl) Joseph Nettis/Photo Researchers; F11 (t) Jeff Schultz/Alaska Stock Images; F16 (t) Buck Ennis/Stock, Boston; (b) J.C. Carton/Bruce Coleman, Inc.; F21 (tl) Michael Holford Photographs; (br) Spencer Grant/PhotoEdit; F25 Marco Cristofori/ The Stock Market; F26 (tr) Corbis; (bl) Shaun Egan/Tony Stone Images; F30-F31 (bg) Jerry Lodriguss/Photo Researchers; F31 (br) Picture PerfectUSA; F32 (b) James M. Mejuto Photography; F34 (bl) Mark E. Gibson; (br) Bob Daemmrich/Stock, Boston; F36 (b) Myrleen Ferguson/PhotoEdit; F37 (b), (tl) Jan Butchofsky/Dave G. Houser; F39 (tr), (c) Richard Megna/Fundamental Photographs; F42 (b) Randy Duchaine/The Stock Market; F44 (b) Tom Skrivan/The Stock Market; F45 (tr) David Woodfall/Tony Stone Images; F47 (t) Roy Morsh/The Stock Market; F48 (l) Ed Eckstein for The Franklin Institute Science Museum; (r) Peter Angelo Simon/The Stock Market; F48-F49 Paul Silverman/ Fundamental Photography; F54-F55 (bg) Superstock; (br) David Madison/Bruce Coleman, Inc.; F58 (cl) John Running/Stock, Boston; (b) David Young-Wolff/PhotoEdit; F59 H. Mark Weidman; F60 (t) D & I McDonald/The Picture Cube; F62 (b) Nasa/The Stock Market; (r) Richard Megna/Fundamental Photographs; F64 (b) Edith G. Haun/Stock, Boston; F70 (b) Amy C. Etra/PhotoEdit; F71 (t) Tony Freeman/PhotoEdit; F72 (b) Dave G. Houser; F74 Webb Chappell; F75 John Lei/Omni-Photo Communications; F76 (tr) NASA/Langley Research Center; (bl) Valder/Tormey/International Stock.

Health Handbook - R15 Palm Beach Post; R19 (tr) Andrew Speilman/Phototake; (c) Martha McBride/Unicorn; (br) Larry West/FPG International; R21 Superstock; R26 (c) Index Stock; R27 (tl) Renne Lynn/ Tony Stone Images; (tr) David Young-Wolff/PhotoEdit.

All Other photographs by Harcourt photographers listed below, © Harcourt:
Weronica Ankarorn, Bartlett Digital Photography, Victoria Bowen, Eric Camdem, Digital Imaging Group, Charles Hodges, Ken Karp, Ken Kinzie, Ed McDonald, Sheri O'Neal, Terry Sinclair.

Illustration Credits - Craig Austin A53; Graham Austin B56; John Butler A42; Rick Courtney A20, A51, B14, B55, C22, C40, C41, C62, C64; Mike Dammer A27, A61, B35, B61, C27, C55, C81, C109, D23, D51, D93, E35, E53, F27, F51, F77; Dennis Davidson D58; John Edwards D9, D17, D18, D31, D37, D39, D40, D59, D64, D68, D69, D70, D71, D76, D77, D78, D80, E17, D10, F45, F60; Wendy Griswold-Smith A37; Lisa Frasier F78; Geosystems C36, C42, C44, C103, D12, D46; Wayne Hovice B28; Tom Powers C8, C13, C20, C34, C50, C102; John Rice D16, B7, B21, B50; Ruttle D36, D7; Rosie Saunders A15; Shough C90, F22, F46, F71, F72.